ORION

**Return to the Moon
Reference Guide
Ronald P. Milione
ISBN: 9798351261218
Series 1**

Contents

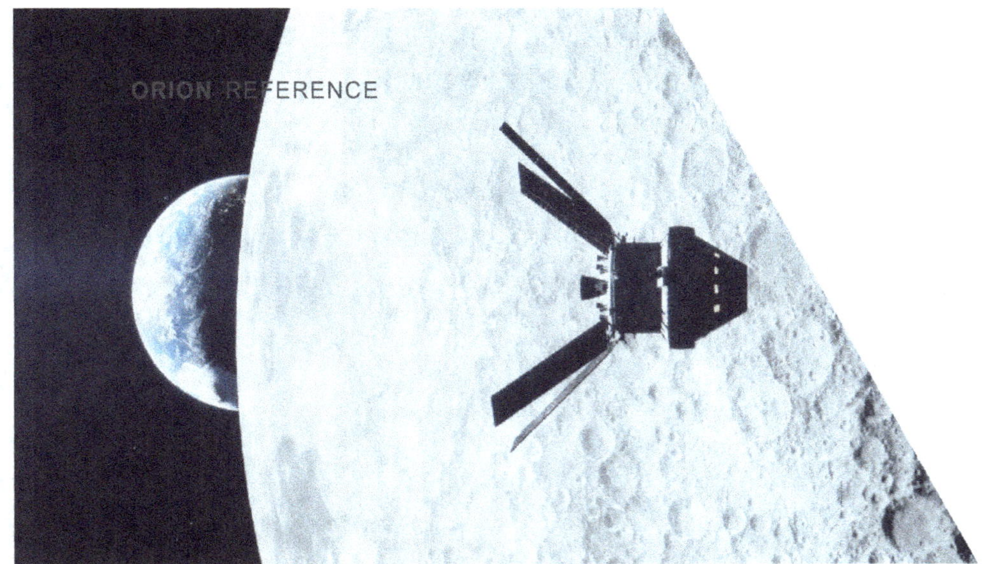
ORION REFERENCE

IV

The Elevator Pitch

Orion and the Artemis Missions

NASA will send astronauts deeper into space than humans have ever gone—40,000 miles beyond the Moon and farther than any of the Apollo missions. In a few short years, the first woman and first person of color will walk on the surface of the Moon and begin to establish long- term exploration for the first time.

But first, to ensure those missions are safe and successful, later this year NASA will launch an uncrewed Orion spacecraft atop the most powerful rocket ever built, the Space Launch System, to test our capabilities to orbit the Moon and return to Earth.

30 seconds

NASA will send astronauts deeper into space than humans have ever gone—40,000 miles beyond the Moon and farther than any of the Apollo missions. In a few short years, the first woman and first person of color will walk on the surface of the Moon.

But first, to ensure those missions are safe and successful, later this year NASA will launch an uncrewed Orion spacecraft atop the most powerful rocket ever built, the Space Launch System, to test our capabilities to orbit the Moon and return to Earth.

America has entered a new era of exploration. We call it Artemis. It will lead humanity back to the Moon, this time sustainably, and prepare us for our next giant leap: exploration of Mars.

Orion is the vehicle that will take astronauts on Artemis missions. It's the only spacecraft capable of human deep space flight and high-speed return to Earth from the vicinity of the Moon. More than just a crew module, Orion has a launch abort system to keep astronauts safe if an emergency happens during launch, and a European-built service module, which is the powerhouse that fuels and propels Orion and keeps astronauts alive with water, oxygen, power, and temperature control.

60 Seconds

You've all heard of the Apollo missions when NASA astronauts first stepped on the Moon. Now, America has entered a new era of exploration. We call it Artemis. It will lead humanity back to the Moon, this time sustainably, and prepare us for our next giant leap: exploration of Mars.

Artemis was the twin sister of Apollo and goddess of the Moon in Greek mythology. Artemis now personifies NASA's path to the Moon and beyond.

To ensure Artemis missions to deep space are safe and successful, later this year during the Artemis I uncrewed mission, NASA will launch the Orion spacecraft atop the most powerful rocket ever built, the Space Launch System, to test our capabilities to orbit the Moon and return to Earth.

During NASA's Artemis II mission, NASA will send astronauts to orbit the Moon, testing out many of the systems Orion will need on future missions.

On Artemis III, the world will witness the first woman and first person of color walking on and exploring the Moon. Orion will deliver astronauts to a commercially built Human Landing System, which will take them to the Moon's surface.

Orion is the vehicle that will take astronauts on Artemis missions. It's the only spacecraft capable of human deep space missions and high-speed return to Earth from the vicinity of the Moon. Orion is versatile, with a reusable and long-duration infrastructure that will last multiple decades and support missions of increasing complexity.

More than just a crew module, Orion has a launch abort system to keep astronauts safe if an emergency happens during launch, and a European-built service module, which is the powerhouse that fuels and propels Orion and keeps astronauts alive with water, oxygen, power, and temperature control.

INTRODUCTION

Artemis

Artemis I will send an uncrewed Orion spacecraft atop the Space Launch System on a journey beyond the Moon. Artemis II will follow, sending a crew around the Moon and back to Earth. In parallel, NASA and its industry partners will build human lunar landers, the final element of the transportation system needed to return humans to the surface of the Moon. Artemis III will mark this historic milestone and the beginning of humans living, learning, and working on the Moon.

A coalition of NASA, industry, and partner nations will establish an enduring human and robotic presence on the Moon in order to meet the agency's science and exploration goals. The Space Launch System will launch the Orion spacecraft to deliver crew to the lunar-orbiting Gateway, where landers will ferry the astronauts to and from the surface of the Moon. NASA and its partners will expand the Gateway's capabilities, allowing for longer stays aboard the orbiting outpost.

The Artemis Base Camp, with a surface habitat and both pressurized and unpressurized rovers, will establish our first foothold on the lunar frontier for extended expedi- tions. Artemis astronauts will advance new technologies while learning more about the Earth, Moon, and our history in the solar system. Together, the Gateway and the Artemis Base Camp will support mission simulations to prepare for the next step – sending astronauts to Mars.

O R I O N

Introduction

The Orion spacecraft is built by NASA and prime contractor Lockheed Martin. It is currently the only spacecraft capable of crewed deep space flight and high- speed return to Earth from the vicinity of the Moon.

NASA's Artemis program is leading the charge to land the first woman and first person of color on the surface of the Moon, and Orion is the spacecraft that will carry astronauts from Earth to the lunar vicinity and return them back home.

Orion is specifically designed to carry astronauts on deep space missions' father than ever before. It will provide protection from solar radiation and high-speed entry into Earth's atmosphere, as well as advanced and reliable

technologies for communication and life support. Orion's missions will span multiple phases, as part of NASA's framework to build a flexible, reusable, and long-duration infrastructure that will last multiple decades and support missions of increasing complexity.

The Moon is the proving ground for sending astronauts to Mars. It gives us opportunities to test new tools, instruments, and equipment to expand humanity into our solar system. Mars remains the horizon goal. NASA expects to take advantage of the unique exploration capabilities of both SLS and Orion and emerging commercial capabilities to build and support the Gateway.

▲   The team at Kennedy Space Center prepares the Artemis I Orion for transport from the Launch Abort System Facility to the Vehicle Assembly Building where it will be stacked atop the Space Launch System on Oct. 18, 2021.

Orion Quick Facts

Performance

Number of crew	4
Mission Duration	up to 21 days

Trans-Lunar Insertion Mass

Artemis I	53,000 lbs.
Artemis II	58,000 lbs.

Gross Liftoff Weight

Artemis I	72,000 lbs.
Artemis II	78,000 lbs.

Height

Crew module + service module	26 ft.
Orion stack (launch abort system + crew module + service module)	67 ft.
SLS Block 1 Configuration (Orion + SLS stack)	322 ft.

Post-Trans Lunar Insertion Mass

Artemis I	**5 1 , 5 0 0** lbs.
Artemis II	**5 7 , 0 0 0** lbs.
Usable Propellant	**1 9 , 0 0 0** lbs.

Total Change in Velocity (ΔV) with Fully Loaded Propellant Tank

Artemis I	53,000 lbs.
Artemis II	58,000 lbs.

Orion Fun Facts

» The Orion spacecraft is built using 355,056 individual parts:

  » 41,858 launch abort system parts

  » 209,544 crew module parts

  » 103,023 service module parts

  » 631 integration parts

» There are 77,150 different types of parts that are used on Orion.

Launch Abort System

Orion's launch abort system (LAS) is designed to carry the crew to safety in the event of an emergency during launch or ascent atop the agency's Space Launch System (SLS) rocket. It can activate within milliseconds to pull the vehicle to safety and position the crew module for a safe landing.

The LAS is divided into two parts: the fairing assembly and the launch abort tower. The fairing assembly is a shell composed of a lightweight composite material that protects the capsule from the heat, wind, and acoustics of the launch, ascent, and abort environments. The launch abort tower includes the system's three motors.

In the case of an emergency, the three solid rocket motors work together to propel the astronauts inside Orion's crew module to safety: the abort motor pulls the crew module away from the launch vehicle; the attitude control motor steers and orients the capsule; then the jettison motor ignites to separate the LAS from the crew module prior to parachute deployment.

During a normal launch, once the SLS clears most of the atmosphere, the job of the LAS is done. The jettison motor fires to separate the LAS from the spacecraft, and from

this point, abort scenarios are handled by the engines in the service module. Fully jettisoning the entire LAS once it's no longer needed frees Orion of thousands of pounds that will make it lighter for its trips into space.

The tower-like abort structure is specifically built for deep space missions and to ride on a high-powered rocket. It is positioned with its motors on top of the Orion crew module and designed to pull the crew module away from the rocket, rather than push it away with motors at the base, as some other escape systems are built to do. This design minimizes the mass that aborts in an emergency by leaving the service module behind.

The abort motor generates 400,000 pounds of thrust, making Orion's LAS the highest acceleration escape system ever tested. The puller-style system with the tower above the spacecraft is the first of its kind capable of controlled orientation after separating from the rocket. In the event of an abort, the LAS can outrun even the SLS rocket, which generates 8.8 million pounds of thrust.

The LAS development is managed by NASA's Langley Research Center and Marshall Space Flight Center.

◀ In this view from above inside the Launch Abort System Facility at NASA's Kennedy Space Center in Florida, work takes place on Aug. 21, 2019, to integrate segments of the launch abort system for Artemis I.

Launch Abort System Quick Facts

Height	50 ft. with ogive panels
	44 ft. without ogive panels
Diameter	3 ft. tower
	17 ft. at base
Liftoff Weight	17,000 lbs.
Propellant Weight	5,700 lbs.

Abort Motor Weight	7,600 lbs.
	includes 4,700 lbs. propellant
	400,000 lbs. thrust
Altitude Motor Weight	1,700 lbs.
	includes 650 lbs. propellant
	7,000 lbs. thrust
Jettison Motor Weight	900 lbs.
	includes 350 lbs. propellant
	40,000 lbs. thrust

▼ The Orion launch abort system and crew module attached to the abort test booster are readied for flight ahead of the Ascent Abort-2 flight test on July 2 ,2019. The successful test demonstrated the ability to carry the crew to safety in case of a mishap during ascent.

Launch Abort System Fun Facts

» The abort motor on Orion's launch abort system (LAS) generates 400,000 pounds of thrust in a matter of millisec- onds to quickly propel the crew module away from danger.

» From a launch pad abort, Orion's launch abort system accelerates from 0 to 500 mph in 2 seconds.

  » This is nearly 6 times faster than the average roller coaster.

  » A 2014 Corvette Stingray goes 0 to 60 mph in 3.8 seconds. It would take the Corvette 32 seconds to get to 500 mph, which is impossible, since it cannot go that fast.

  » At take-off, a Boeing 747 is going approximately 225 mph and requires at least 10 seconds to get to that speed – its fastest cruising speed is 570 mph.

  » This is nearly 3 times faster than the speed of an aircraft carrier catapult (a catapult can loft 45,000 pounds from 0 to 165 mph in 2 seconds).

» The LAS solid rocket motors use propellant that burns at a temperature of 4,456°F, nearly half the temperature on the surface of the Sun. While that temperature is hot enough to boil steel, special insulation inside the motor protects the steel case so well that the outside of the case reaches only about 130°F and the crew
cabin remains cool.

» The motors of the LAS use a propellant is a solid rubberlike substance of fuel and oxidizer that has the consistency of a hard rubber eraser.

» Combustion gases exiting the abort motor nozzles travel at a speed of 2,600 mph, more than 3 times the speed of sound and 2 times the speed of a bullet shot from a rifle.

» If the abort motor energy could be converted to electric power, it would be enough to power 13,000 houses for an entire day.

◄ Teams with NASA's Exploration Ground Systems (EGS) and contractor Jacobs integrate the launch abort system (LAS) with the Orion spacecraft inside the Launch Abort System Facil ty at NASA's Kennedy Space Center in Florida on July 23, 2021.

▲ The NASA insignia, also called the "meatball," and the American Flag are applied to the Orion crew module back shell for the Artemis I mission on Oct. 28, 2020, inside the Neil Armstrong Operations and Checkout Building at NASA's Kennedy Space Center in Florida.

Crew Module

The crew module is the pressurized part of the Orion spacecraft, sometimes referred to as the capsule, where crew will live and work on their journey to the Moon and back. The crew module can accommodate four crew members for up to 21 days, and provides a safe habitat through launch, on-orbit operations, landing, and recovery. The crew module contains advances in life support, avionics, power systems, and advanced manufacturing techniques, and is the only portion of Orion that returns to Earth at the end of flights.

Inside the crew module, environmental control and life support systems maintain cabin temperature, pressure, humidity, oxygen, and carbon dioxide levels, keeping crew

healthy and comfortable on long missions. A cockpit with glass displays allows full control of Orion by the crew. The sealed capsule also provides radiation protection needed to safeguard crew and spacecraft systems from cosmic and solar radiation seen in deep space, and micrometeoroid protection from items found in the space environment.

On missions to the Moon, the crew module will be home to the astronauts. Water tanks and a dispenser provide drinking water and a simple way to rehydrate and warm food. The crew module's waste management system, or lavatory, is designed for multi-week missions, privately accommodating both sexes while in zero gravity. Astronauts will also have to exercise, and Orion's crew module has a built-in exercise device that provides both aerobic and strength training. The environmental control system even removes excess heat, humidity, and odor during exercise.

The underlying structure of the crew module is called the pressure vessel. The pressure vessel consists of seven large aluminum alloy pieces that are joined together using friction-stir welding to produce a strong, yet light-weight, air-tight capsule. The seven major structural pieces include the barrel, tunnel, forward and aft bulkheads, and three cone panels. Original designs had 33 welded pieces, but the Artemis I crew module and beyond have seven welded pieces, improving manufacturability and saving more than 700 pounds of excess weight.

The floor structure where the crew seats are attached and where the crew stowage lockers are located is called the backbone assembly, which is a nine-piece bolted structure of crisscrossing beams. The backbone, made of aluminum, also provides additional structural support for the crew module. Most of the equipment the crew needs to live in

space, such as food, medical kits, emergency equipment such as masks and fire extinguisher, sleeping bags, and the pressure suits worn for launch and return to Earth are stored in lockers located here.

Covering the pressure vessel is the protective cover on the sides of the crew module known as the backshell, made up of 1,300 thermal protection system tiles. The tiles are made of a silica fiber material similar to those used for more than 30 years on the space shuttle. The tiles will protect the spacecraft from both the coldness of space and the extreme heat of entry into Earth's atmosphere.

The bottom of the Orion capsule, which will be pointed into the heat as Orion returns to Earth, is covered by the world's largest ablative heat shield, measuring 16.5 feet in diameter. The heat shield will protect Orion as it enters Earth's atmosphere travelling about 25,000 mph and endure temperatures of nearly 5,000 degrees Fahrenheit
— about as half as hot as the Sun. The outer surface of the heat shield is made of 186 billets, or blocks, of an ablative material called Avcoat, a reformulated version of

The Orion crew module for Artemis I is lifted by crane on July 16, 2019, in the high bay inside the Neil Armstrong Operations and Checkout Building high bay at NASA's Kennedy Space Center in Florida.

the material used on the Apollo capsules. The Avcoat is bonded to a titanium skeleton and composite skin that gives the shield its shape and provide structural support for the crew module during descent and splashdown. During descent, the Avcoat ablates, or burns off in a controlled fashion, transporting heat away from Orion.

The crew module includes a number of mechanisms that provide the way to enter and leave the module through hatches. It also includes mechanicsm that provide docking capability for on-orbit activities with other spacecraft, as well as as mechanisms for the separation or jettison functions from other parts of Orion when the crew module needs to separate for entry and landing.

While the service module functions as the main propulsion system for Orion, the crew module has a propulsion system comprised of 12 small engines called reaction control system (RCS) thrusters that provide full control of module translation and rotation both on orbit and during entry. When the crew module separates from the service module for entry into Earth's atmosphere, the 12 RCS thrusters control the spacecraft's return.

The capsule also houses Orion's parachute system that ensures a safe landing for astronauts returning from deep space missions to Earth. The system includes a series of 11 parachutes that are deployed in a sequence to slow down the module from speeds exceeding 25,000 mph to 20 mph or less and provide a safe speed for splashdown at landing.

Crew modules for Artemis missions are built by prime contractor Lockheed Martin, and primary structure welding takes place at NASA's Michoud Assembly Facility in New Orleans. Orion piece parts and components are provided by hundreds of suppliers across the nation, and parts are delivered to the assembly facility located at the Kennedy Space Center. Orion is then outfitted with all these parts and internal components, assembled by teams at Kennedy, then transferred to the Exploration Ground Systems (EGS) team which performs final preparations on the spacecraft for launch.

The Orion crew module for Artemis I undergoes a direct field acoustics test, where it was exposed to maximum acoustic levels that the vehicle will experience in space.

Crew Module Quick Facts

Height	11 ft.
Diameter	16.5 ft.
Habitable Volume	316 cu. ft.
Pressurized Volume	220 lbs
Lunar Payload Return	220 lbs.
Artemis I Liftoff Weight	20,500 lbs.
Artemis II Liftoff Weight	22,900 lbs.
Artemis I Landing Weight	18,200 lbs.
Artemis II Landing Weight	20,500 lbs.
Engines/Thrusters	12 Reaction Control System (RCS) Thrusters 160 lbs. of each thrust

Crew Module Fun Facts

» From an anthropomorphic standpoint, Orion is designed to accommodate 99 percent of the human population, which is a larger range than every other NASA or Department of Defense project past, present, or future. The spacecraft can accommodate astronauts as small as a 4-foot-10-inch tall female and as tall as a 6-foot-5-inch male.

» Orion's crew module provides about 30 percent more cubic volume of space than the Apollo capsule.

» Orion can sustain a crew of four for up to 21 days in orbit during its initial missions to low lunar orbit. Missions to more distant destinations may have a crew of 2 to allow more room for consumables onboard.

» Orion can orbit the Moon autonomously without any crew onboard for over six months.

The Artemis I Orion pressure vessel, which is the underlying structure of the crew module, arrived at the Neil Armstrong Operations & Checkout Building at NASA's Kennedy Space Center in Florida in early February 2016 and was maneuvered into a work stand.

The Orion crew and service module stack for Artemis I is lifted out of the Final Assembly and Test (FAST) cell inside the Neil Armstrong Operations & Checkout Building at NASA's Kennedy Space Center in Florida.

European Service Module

Orion's European Service Module is the powerhouse that fuels and propels the Orion spacecraft in space. Located below the crew module, it provides critical functions including propulsion, thermal control, and electrical power generated by solar arrays. The service module will also provide life support systems including water, oxygen, and nitrogen for the crew. In addition to its function as the main propulsion system for Orion, it is responsible for orbital maneuvering and position control. The cylindrical module is unpressurised and about 13 feet high, including the main engine and tanks for gas and propellant, and is designed for long duration missions to deep space destinations.

During launch, the service module fits into a 17-foot-diameter housing, surrounded by three spacecraft adapter jettison fairing panels. The panels offer protection from harsh environments such as heat, wind, and acoustics during launch. Once Orion is above the atmosphere, the fairing panels surrounding the service module are jettisoned, and the solar arrays unfold to span about 62 feet. Following launch, Orion's service module will propel the spacecraft on its mission and help it return to Earth, detaching before the crew module enters Earth's atmosphere.

To push Orion to its destination, the service module is equipped with a total of 33 engines: one main engine (NASA-provided refurbished Orbital Maneuvering System engine (OMS-E); eight auxiliary engines (NASA- provided Aerojet R4D-11), and 24 reaction control thrusters (ESA-provided engines that are the same model as those used on the Automated Transfer Vehicle (ATV). These three types of engines provide the propulsion for lunar orbit injection and return to Earth, as well as attitude control for the crew module. The propulsion system can also be used, during some late phases of the launch, for crew safeguard during potential abort scenarios.

Built by Thales Alenia Space in Italy, the primary structure of the service module is the backbone of the entire vehicle. Like the chassis of a car, the structure holds everything together. It is covered with Kevlar to absorb shocks from micrometeorites and debris impacts.

During launch, the service module is held in place by the held in place by the spacecraft adapter. The spacecraft adapter connects to the Orion stage adapter that attaches Orion to the SLS rocket.

A crew module adapter connects the capsule to the service module. The crew module adapter houses electronic equipment for communications, power, and control, and includes an umbilical connector that bridges the electrical, data, and fluid systems between the main modules.

Computers control all aspects of the service module. Avionics manage the powered equipment and the data exchange services based on instructions received from Orion's flight computers in the crew module. Nearly seven miles of cables send commands and receive information from sensors.

The service module's electrical power system provides the power for the entire spacecraft. The system manages the power generated by the four solar array wings of the module. Each wing is made of three panels, and each panel is approximately 6.5 x 6.5 feet (2 x 2 meters). The total length of each wing is nearly 23 feet (7 meters). There are a total of 15,000 gallium arsenide cells on the four arrays used to convert light into electricity, and the arrays can turn on two axes to remain aligned with the Sun for maximum power. A power control and distribution unit provide the power interface between the service module and the crew module adapter, distributes electrical power to the service module's electrical equipment, and protects the power lines. The module's thermal control system includes radiators and heat exchangers to keep the astronauts and equipment at a comfortable temperature. The thermal control system includes an active portion, which transfers the heat of the entire vehicle to the service module radiators, and a passive portion, which protects the module from internal and external thermal environments.

A consumable storage system provides potable water, nitrogen, and oxygen to the crew module. Potable water is provided by the water delivery system and stored in four metal bellow tanks, covering usable water needs of the crew for the duration of the mission. Oxygen and nitrogen are provided by the gas delivery system and stored in four tanks, the allocation of the tank to each gas being mission dependent.

The service module is built by main contractor Airbus under a contract for ESA (European Space Agency), with many companies across 10 European countries as well as the United States supplying components. Airbus has drawn on its experience as prime contractor for the ATV, which provided supplies to the International Space Station. The final product is assembled in Europe before being shipped to NASA's Kennedy Space Center, where it is integrated with the rest of the Orion spacecraft.

Service Module Quick Facts

Length	15.7 ft.
Diameter	16.5 ft.
Artemis I Weight	33,700 lbs.
Artemis II Weight	34,400 lbs.
Engine/Thrusters	1 Orbital Maneuvering System; 6,000 lbs. of thrust
	8 Auxiliary Engines; 110 lbs. of each thrust
	24 Reaction Control System Thrusters; 50 lbs. of thrust each
Solar Arrays	4 arrays
	15,000 solar cells
	62 ft. when deployed
	11kW regenerable electrical power

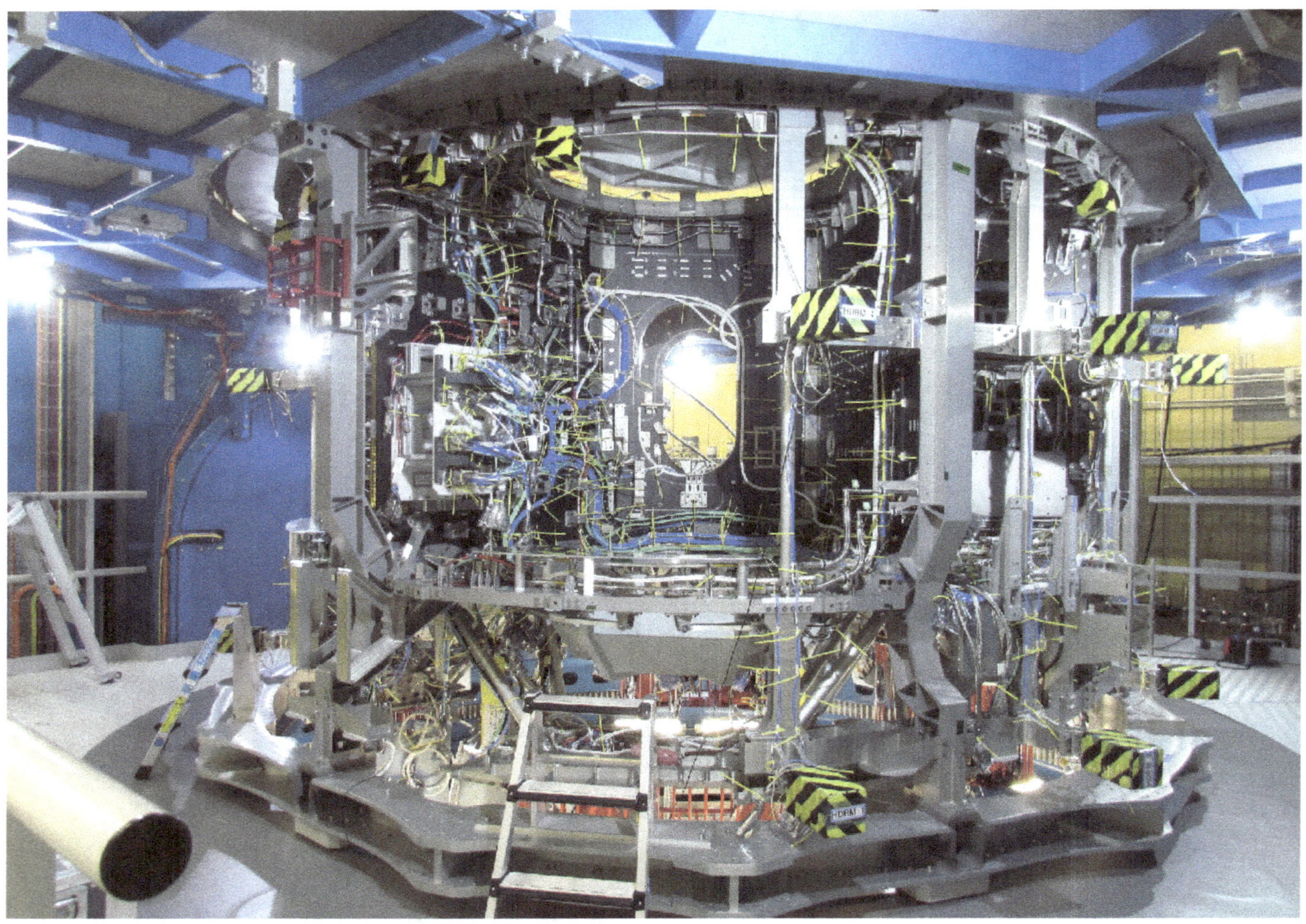

▲ The European Service Module that will be used on Artemis II, the first Artemis mission with crew, is shown in assembly at Airbus in Bremen, Germany on July 1, 2019.

Service Module Fun Facts

» The European Service Module is comprised of more than 20,000 parts and components that must fit together perfectly and perform reliably.

» The computer in Orion's service module controls 33 engines and reads over 100 pressure and temperature sensors.

» 1,000 sensors collected data on Orion's service module as it was vigorously shaken on a vibration table at NASA Glenn Research Center's Neil A. Armstrong Test Facility in Sandusky, Ohio. This testing simulated how Orion's structure flexed and stood up to 35 tons of spacecraft weight during a launch.

» Each of Orion's four solar array wings are made of three panels that provide enough electricity to power two three-bedroom homes.

» In space, Orion's orbital maneuvering system engine, called the OMS-e, produces 6,000 pounds, or 26.7 kN, of thrust. Here on Earth, if you were to strap an OMS-e to a car, it would accelerate from 0-60 mph in approximately 2.7 seconds, ranking it among the world's top 10 fastest production cars.

» Orion's service module main engine for Artemis I has been to space 19 times on three different space shuttles.

Avionics

The Orion spacecraft houses a number of state-of-the-art avionics units to handle data generated by on-board systems, control the various functions of the spacecraft, execute commands sent from Earth or by the crew, and return systems telemetry for insight into systems status.

The avionics and other electronics used in Orion are almost entirely driven by software and commercial processor technologies that have been ruggedized to endure extreme radiation and temperature fluctuations. Orion's updated avionics also can handle the severe acoustic and vibration environments associated with launch, orbit, a fiery entry into Earth's atmosphere, and a saltwater landing.

Orion's avionics system consists of seven main subsystems: command and data handling; guidance, navigation, and control (GN&C); communications and tracking; displays and controls; instrumentation; power; and flight software (FSW).

Command and Data Handling

The brains of the Orion spacecraft consist of two vehicle management computers (VMC), manufactured by Honeywell. A VMC is a robust system that delivers more computing power to the Orion spacecraft than any previous crewed spacecraft.

Each of the two VMCs is made up of two flight computer modules (FCM), which are in charge of executing flight control and other vehicle software, a display control module for astronaut displays, and a communication control module that allows commands and data to flow between Orion and mission control. The FCMs provide a high integrity platform to house software applications and have sufficient processing power to perform command and control of Orion.

Each of the four FCMs are internally redundant and continually check all operations to be sure they match. If the FCMs ever detect a difference between them, due to a hardware failure or a radiation upset, the different FCM 'fails silent' by stopping all outputs so that a potentially corrupted FCM doesn't issue critical

commands to the vehicle. The FCM then resets itself, listens to the other FCMs to relearn where the vehicle is and what is happening, and then rejoins the other FCMs in controlling the vehicle – all within 22 seconds. Having four FCMs on the vehicle allows the flight software to continue firing thrusters and flying as Orion transitions through the challenging radiation environment of the Van Allen belts, since the odds of all four FCMs being upset in the same 22 seconds are extremely low.

The four redundant FCMs greatly improve system reliability, yet Orion adds another measure of backup capability with the addition of a completely different computer, capable of running different code should it ever be needed. In the unlikely event that something goes wrong with the primary flight computers on Orion, a dissimilar processing platform with dissimilar flight software is hosted on the vision processing unit (VPU). This dissimilar computer and software provide a backup function to the redundant FCMs during critical phases of flight, with a focus on crew survival and return functions in the highly unlikely scenario where anything renders all the FCMs ineffective. The VPU also provides a place to store data during times when Orion can't communicate with the ground.

Eight power and data units (PDU) connect the flight computers and the software to the rest of Orion. These PDUs control the power to every component on the vehicle and they control effectors such as valves, thrusters, and heaters. All sensor data such as temperature and pressure are routed through the PDUs as well. The PDUs also play a critical role in communicating with the rocket that launches Orion and puts it on its trajectory to the Moon.

Orion's onboard data network is a triple redundant network that allows the FCMs to communicate with all of the other avionic components on Orion. It uses a networking technology called Time-Triggered Gigabit Ethernet (TT-GbE) that is capable of moving data at a rate 1,000 times faster than systems used on the shuttle and space station. The TT-Gbe allows NASA engineers to categorize different types of data and prioritize how it should travel through the onboard network. Time- critical data relating to vital systems like navigation and life support, called time-triggered data, has guaranteed bandwidth and message timing to ensure it is always delivered exactly on time. Data that is critical for delivery but not timing, such as file transfers, is called rate- constrained data and is sent immediately whenever time- triggered data is not present. Data used for non-critical tasks such as crew videoconferencing is delivered over the remaining bandwidth. The technology means that critical data and non-essential data can travel safely over a single network onboard a spacecraft for the first time. It is built upon a reliable commercial data bus that has been hardened to be resilient to space radiation and proven on the Orion Exploration Flight Test-1 (EFT-1). The data system interfaces with all components, including the European Service Module, through radiation hardened network switches.

Guidance, Navigation, and Control

The guidance, navigation, and control (GN&C) system is responsible for always knowing where the spacecraft is and where it is going and controls the propulsion system to keep Orion pointed in the proper direction and on the correct trajectory.

At the center of this system is the GN&C flight software (FSW) that executes on the VMCs. This software receives inputs from navigation sensors and pilot controls and commands the appropriate effectors on the crew module, service module, and launch abort system to accomplish mission objectives. The Orion GN&C software operates across a variety of mission phases, including pre-launch, ascent, Earth orbit, transit to and from the Moon, loiter, rendezvous, docking, entry, and various abort scenarios. The software must operate in both manual and automated modes, and must handle commands from the crew and the ground.

The software must also execute complex guidance and navigation algorithms while controlling highly dynamic configurations during entry, ascent aborts, and orbital maneuvers. The resulting range of algorithm types drives a multi-rate architecture to meet central processing unit usage allocations.

The onboard navigation system for Orion is comprised of a number of redundant sensors for measuring Orion's position (where Orion is in space) and attitude (what direction the spacecraft is pointed). Like most systems on the spacecraft, there are usually at least two of each sensor to increase reliability of the overall system. Because the spacecraft operates in the atmosphere during ascent and entry, in low-Earth orbit, and near the Moon, several different types of sensors are needed. These include:

Orion Inertial Measurement Units (OIMU): Each OIMU contains three rate gyros which measure vehicle body rotation rates and three accelerometers to measure vehicle body accelerations. This inertial data is used by the VMC for onboard navigation to compute spacecraft position, velocity, and attitude.

Global Positioning System (GPS) Receivers: Like ground-based GPS, except these are capable of operating at the very high velocities of spaceflight. The GPS sensor system provides position and velocity updates during low-Earth orbit operations, ascent, and entry. Because the GPS satellites point toward the Earth, GPS can't be used out near the Moon. GPS-based altitude values are the primary triggers for entry events.

Deep Space Network (DSN): This is not an on-board system, but instead a ground-based network of large tracking dishes all around the world that together can determine Orion's location while it is in deep space outside the range of GPS. It relies on very sensitive timing of measurements of communication signals that pass between Orion and the dishes on the ground.

Barometric Altimeter (BALT) Assembly: By sensing the atmospheric air pressure outside the spacecraft during ascent and entry, these assemblies can measure Orion's altitude. They provide a back-up altitude value for parachute and other deployments during entry.

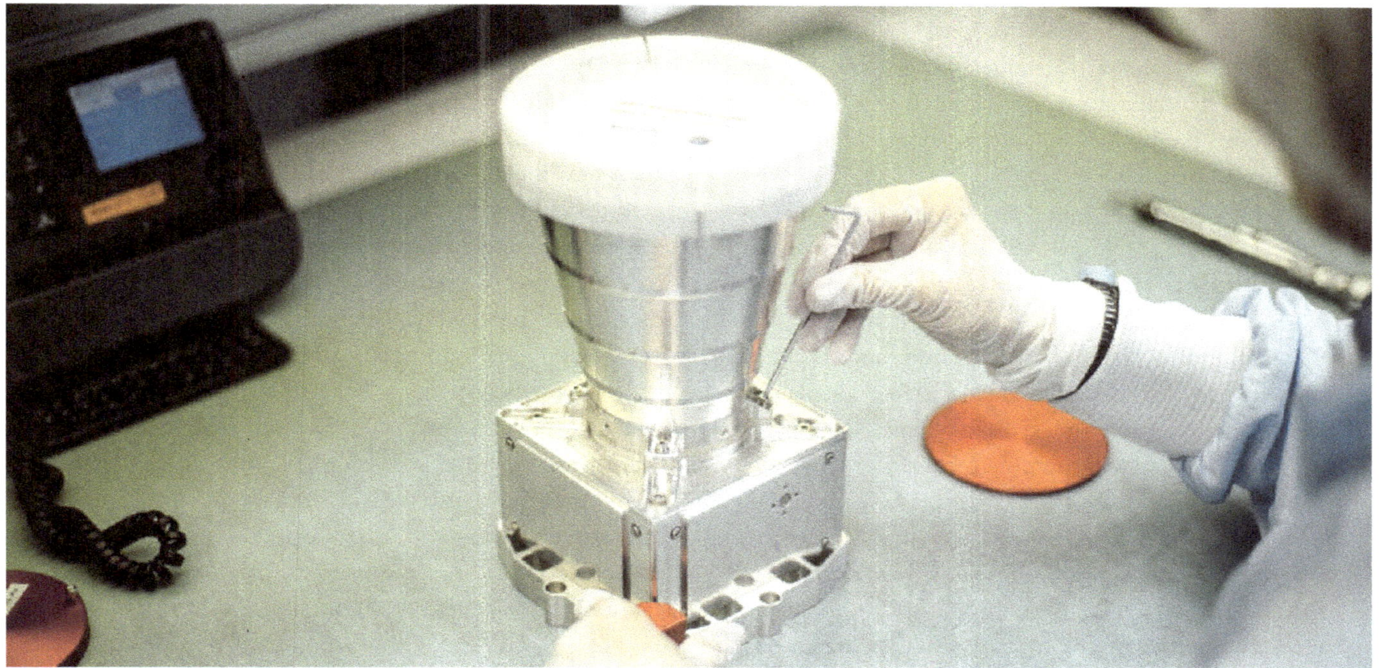

Star Trackers (STR): The star tracker operates like a camera but is very sensitive and takes pictures only of stars. By comparing the pictures to a known star catalog, the sensor determines spacecraft attitude during orbital operations.

Optical Navigation (OpNav) Camera: The OpNav camera takes images of the Moon and the Earth. By looking at the size and position of these objects in the picture, OpNav can determine Orion's range and bearing to that object. OpNav is part of the Orion emergency return system to autonomously operate the vehicle in the event of lost communication with Earth.

Sun Sensors: The Sun sensors are located on the ESA- provided service module and are used to determine the direction of the Sun during emergency safe mode. Knowing where the Sun is ensures that Orion can point its solar arrays in the right direction to keep power flowing.

Communications and Tracking

Orion uses a high-speed communications system, employing four phased array antennas on the crew module and two phased array antennas on the service module. Phased array antennas allow signals to be controlled and directed without requiring any physical movement of the antenna. These will be used for video, data, and voice communications with the spacecraft, as well as command uplink and telemetry downlink to ground stations, NASA's Tracking and Data Relay Satellite Systems, and NASA's Deep Space Network after leaving Earth's orbit. The crew will also have an emergency radio system to allow voice communications anytime during a mission if the main communications system were to fail. After landing, the crew has search and rescue radios as well as satellite phones for communication with recovery forces. There is also an audio system on board that allows astronauts to speak with each other and with the ground while they are wearing spacesuits.

Orion also has a number of cameras inside and outside of the vehicle to help the crew with tasks like docking, interviews and public outreach events, medical conferences, and to help gather engineering pictures and video to make sure the spacecraft is performing properly.

Each of the solar array wings has a wireless camera near the tip that can be pointed to inspect the exterior of the spacecraft and to provide some views that are out of this world.

Displays and Controls

The Orion displays and control equipment is the crew interface to the Orion systems. The displays and controls consist of three display units, seven switch interface panels, two rotational hand controllers, two translational hand controllers, and two cursor control devices. The switch panels and hand controllers' hardware interfaces through serial interfaces to the power and data units, then via the onboard data network to either flight control modules or the display units for processing. The display units utilize a variety of display formats to provide data to the crew for awareness and action when necessary.

The Orion displays and controls are designed for an intensive amount of crew interaction both in nominal and off-nominal scenarios. Electronic procedures have been developed for Orion that allow direct interaction with the display formats enabling reduced workload on the crew. The electronic procedures efficiently step the crew through planned tasks and reduce crew workload by highlighting various telemetry on a display format or queuing up commands. Additionally, the electronic procedures have built in links to the onboard caution and warning system which alert the crew when onboard faults and anomalies occur. The electronic procedures link provides the ability for the crew to bring up electronic procedures which communicate the urgent actions the crew need to take to address the caution and warning condition.

▼ Spacesuit engineers demonstrate how four crew members would be arranged for launch inside the Orion spacecraft, using a mockup of the vehicle at Johnson Space Center in Houston.

Power

The Orion power system can generate and supplying all of the power that is required for its on-orbit operations. The four Orion solar arrays are located on the European Service Module, generate about 11 kilowatts of power and spread 62 feet when extended. Orion's four main batteries are located on the crew module and use small cell packaging technology to ensure crew safety while providing 120 volts of power to the many systems on Orion. The batteries are fully charged before launch to operate the spacecraft until the solar arrays can be deployed once in orbit. The batteries also operate the spacecraft when the solar arrays cannot be pointed at the Sun or in the shadow of the Earth or the Moon. The solar arrays are jettisoned right before entering the Earth's atmosphere, so the batteries also provide all the power needed to keep the astronauts safe for return to Earth and up to 24 hours after splashdown. Power is transferred between the solar arrays and batteries and to the end item loads via the power and data units.

Instrumentation

Accomplishing flight test objectives for the flight tests requires a dedicated instrumentation system that will measure dynamic response of all Orion subsystem performance during critical phases of the Artemis missions. These include structural response and thermal protection system response during entry into Earth's atmosphere. A suite of avionics data acquisition system electronics along with associated cabling supports the large number of channels needed.

The developmental flight instrumentation (DFI) data system is required to measure unique subsystems performance during all phases of flight from launch, cruise, and return to Earth. The DFI diagnostic instrumentation system is required to measure the response of newly designed components and structures to verify and validate engineering models that will be used to predict their future performance.

The architecture of the DFI system is very robust and relies on proven hardware and software to deliver high reliability. The central components are data acquisition units which have two interfaces: one for the sensor interface, and one for the control interface. The sensor interface communicates with the temperature, strain, accelerometers, and acoustic sensors. The control interface in turn communicates with the power, control, recording, telemetry, and time-sync hardware. The sensors can be changed between flights to allow engineers to adjust based on what is learned about a previous flight.

▲   In preparation for installation on the Artemis I spacecraft, technicians extend one of the Artemis I solar array wings for inspection inside the Neil Armstrong Operations and Checkout Building high bay at NASA's Kennedy Space Center in Florida on Sept. 10, 2020, to confirm that it unfurled properly and all of the mechanisms functioned as expected.

Avionics Fun Facts

» The Orion S-Band communications system provides functionality of 4 radios in one, allowing communication with:

   » Multiple space vehicles

   » NASA's Near Space Network Direct-To-Earth Services

   » NASA's Tracking and Data Relay Satellite (TDRS) Services

   » NASA's Deep Space Network (DSN)

   » Any combination of the above

» Orion's glass cockpit provides fully redundant crew controls and displays with over 60 graphical user interface, or GUI formats and interactive electronic procedures - a first in spacecraft history. Instead of relying on physical switches distributed across a vast flight deck, an astronaut will be able to control all of the vehicle systems from a single operator station using 'virtual' switches displayed on GUIs.

» Compared to Apollo's single flight computer, Orion has two simultaneously operating redundant flight computers. Each of these includes two redundant computer modules, giving it a total of four redundant systems.

» One of Orion's two redundant computers is only three quarters the weight of the sole computer aboard Apollo, has 128,000 times more memory and is 20,000 times faster.

Heat Shield

The heat shield is one of the most critical elements of the Orion spacecraft, protecting Orion and the astronauts as they enter Earth's atmosphere travelling about 25,000 mph and reaching temperatures of nearly 5,000 degrees Fahrenheit — about as half as hot as the Sun.

Measuring 16.5 feet in diameter, it's the largest ablative heat shield in the world. The outer surface is made of blocks of an ablative material called Avcoat, a reformulated version of the material used on the Apollo capsules. During descent, the Avcoat ablates, or burns off in a controlled fashion, transporting heat away from Orion.

The Avcoat is first made into large blocks at NASA's Michoud Assembly Facility, then shipped to NASA's Kennedy Space Center in Florida and machined into 186 unique shapes before being applied onto the heat shield's underlying titanium skeleton and carbon fiber skin. Engineers conduct non-destructive evaluations to look for voids in the bond lines, as well as measure the steps and gaps between the blocks. The gaps are filled with an adhesive material and then reassessed.

After the thermal protection system has been applied and inspected, engineers and technicians put the heat shield through a thermal cycle test in the high bay of the Operations and Checkout (O&C) Building at Kennedy. The thermal cycle test ensures the thermal protection

blocks are properly bonded and will perform as expected when they are exposed to the extreme temperatures during the mission.

The heat shield is then given a coat of white epoxy paint. Aluminized tape is applied after the painted surface dries to dissipate electrical surface charges and maintain acceptable temperatures. Once all testing has been completed, the heat shield is bolted to the crew module.

The Orion heat shield was redesigned from a single piece system to individual blocks of material following Exploration Flight Test-1 in 2014. Before the time-saving block system was used, a fiberglass-phenolic honeycomb structure was bonded to the structure's skin. Then each of the 320,000 tiny honeycomb cells were individually filled with Avcoat by hand, inspected by X-ray, cured in a large oven, and robotically machined to meet precise thickness requirements.

This new design introduced several considerations that prompted further testing for risk reduction. Engineers performed more than 30 tests across the United States on the new design to investigate the effects of the block structure that could disrupt the smooth airflow and cause localized heating spots. Understanding both effects is critical to confirm the heat shield will thermally protect the astronauts during entry into Earth's atmosphere.

Inside the Neil Armstrong Operations and Checkout Building high bay at NASA's Kennedy Space Center in Florida, technicians with ASRC Federal inspect AVCOAT block bonding on the Artemis II heat shield on July 2, 2020.

The Avcoat material was tested at the NASA Ames Research Center Arc Jet Complex and at the NASA Johnson Space Center's Atmospheric Reentry Materials and Structures Evaluation Facility through its closing in 2014. The Arc Jet facilities were used to simulate the heating and airflow conditions that occur during entry. Thermal testing also has been performed at Johnson Space Center's Radiant Heat Test Facility (RHTF) periodically from the onset of the Orion Program up until March 2021. During these tests, the Avcoat surface reached temperatures of over 3,000 degrees Fahrenheit. Heat shield testing concluded in March 2019 at NASA's Langley Research Center with a six-inch Orion heat shield model in the 20-inch Mach 6 wind tunnel. The six-inch Orion heat shield model was machined to represent small-scale features, including the patterns expected as the heat shield ablates during return to Earth.

Backshell panels

Orion's backshell panels and forward bay cover will also provide protection from the excessive heat of entry into Earth's atmosphere, as well as the micro-meteoroid debris which could be encountered during missions. There are 1,300 thermal protection system tiles made of a silica fiber material called AETB-8 (Alumina Enhanced Thermal Barrier). The tiles are similar to those used for more than 30 years on the space shuttle, and incorporate a stronger coating called "toughened uni-piece fibrous insulation," or TUFI coating, which was used toward the end of the Space Shuttle Program. On average, the tiles are 8-inches by 8-inches and many are standard in size allowing them to have the same dimensions with the same part number.

Heat shield back shell panels are prefitted on the Orion spacecraft inside the Neil Armstrong Operations and Checkout Building high bay at NASA's Kennedy Space Center in Florida on Feb. 1, 2018.

Heat Shield Fun Facts

» During return to Earth from a mission to the Moon, Orion and its heat shield must protect the vehicle and crew from external temperatures up to around 5,000°F.

  » This is hotter than the melting point for titanium, which is about 3,000°F.

  » It's also about half the surface temperature of the Sun, which is about 10,000°F.

» While the outside temperatures during entry into Earth's atmosphere will reach 5,000°F, inside the spacecraft it will be in the mid-70s – hotter than molten lava on the outside and cool as a cucumber on the inside.

» The heat shield on the bottom of the crew module is 16.5 feet wide. It is the largest ablative heat shield in the world. Orion's thermal protection system is one of the most important parts of the spacecraft and is responsible for protecting the astronauts during their return.

Crew Systems

Orion has been specially designed and tested to make sure astronauts can live and work safely in the capsule during multi-week Artemis missions. Countless details inside and outside the spacecraft are meticulously thought through to ensure that astronauts are protected during anything they may encounter. From life support to spacesuits to crew accommodations, engineers have worked hard to evaluate and improve Orion's features so that astronauts can complete missions in deep space comfortably, efficiently, and safely.

Environmental Control and Life Support Systems

On Orion, environmental control and life support systems (ECLSS) will make the crew module a habitable, safe place for astronauts and is key to survival as they travel on deep space missions.

On the International Space Station, a safe return to Earth is only hours away, but that is not the case for deep space missions. Orion provides an ECLSS that is fault tolerant and less complex than previous NASA human spacecraft. The key components of the ECLSS include atmosphere revitalization, pressure control, crew water supply, and crew waste management. For Orion, these systems must be mass and volume efficient as well as dependable.

Atmosphere revitalization is the highest priority on deep space missions. Systems must not only provide oxygen and remove carbon dioxide from the atmosphere, but also prevent gases like ammonia and acetone, which humans emit in small quantities, from accumulating. They also must provide adequate ventilation for the crew and filter particles and microbes. Orion has a new carbon dioxide and humidity removal system which is regenerable, a key for saving mass and volume on deep space vehicles. The system, when exposed to the cabin air, absorbs carbon

dioxide and humidity. When exposed to the vacuum of space, the carbon dioxide and humidity are vented overboard, and the system is regenerated back to a clean state to return to cleaning the cabin air. On other human spacecraft such as the space shuttle, a method using expendable chemicals was used to remove carbon dioxide. For perspective, these chemicals took up the volume of nearly 143 basketballs. Orion's system will take up the space of only 16 basketballs and save over 100 pounds.

Orion uses high pressure oxygen and nitrogen tanks to provide the pressure control for the crew environment. Using these tanks simplifies the system to provide for greater reliability on deep space missions. The pressure control system can be manually operated by the crew if required in a severe situation.

A water supply system stores and distributes potable water to the crew for drinking, food preparation, and medical and hygiene needs.

Environmental monitoring maintains the spacecraft's temperature, humidity and pressure, and detects when the spacecraft's enclosed environment is compromised, causing it to become unsafe. For deep space missions, Orion has an integrated crew survival capability, using pressurized suits for several days in order to return the crew home safely in the event of a total cabin environment loss due to pressure or air contamination. Not since Apollo has this capability been required.

The ECLSS on Orion balance between the ever-present constraints of launch mass, volume, system fault tolerance, and reliability of resources to sustain astronauts and keep them safe in deep space.

Universal Waste Management System

Orion's crew module has a new space toilet, called the Universal Waste Management System (UWMS), making the essential task of going to the bathroom easier for both women and men and reducing the ever-important mass and volume of the system to be launched into deep space.

Looking back, astronauts on Mercury, Gemini, and Apollo did not have toilets. They urinated into diapers and bags and brought their solid waste mixed with bactericide in bags back home. Skylab was the first American spacecraft with a toilet, followed by the space shuttle. The space shuttle toilet was a full size larger (about 12.3 cubic feet) and a massive system, using several separate motors and fans for operations.

Orion's toilet operates in a similar way, using air flow to pull fluid and solid waste away from the body and into the proper containers, but is improved for the needed mass and volume constraints of deep space flight in Orion, and is more accommodating to female astronauts. Based on their input, the shape of the seat for solid waste and design of the funnel for urine has been changed, and they can be used simultaneously.

It's also self-contained and compact at about 5 cubic feet in volume, thus approximately 60 percent smaller and lighter than the space shuttle toilet, easier to use, and built to increase crew comfort. A new automatic air flow feature helps with odor control, fewer control interfaces simplifies crew operations, and a more ergonomic design requires less clean-up and maintenance time, as do corrosion- resistant, durable parts.

Pre-treated urine, which prevents the generation of ammonia from the breakdown of the urine, is stored in a urine tank and then vented overboard each day by the crew much like the space shuttle. Solid waste is collected in fecal canisters, which the crew replace every few days, and stored in Orion for the 21-day mission. The canisters have filtered caps to control odor and gas build up generated within the cans.

What makes the UWMS universal is that it can easily integrate into different spacecraft and systems, keeping our astronauts healthy and safe as they live, work, and learn on Orion farther from Earth than ever before.

▼ The Universal Waste Management System (UWMS) that will fly on NASA's Artemis II mission is self-contained and compact at about 5 cubic feet in volume, making it 60 percent smaller and lighter than the space shuttle toilet, easier to use, and built to increase crew comfort.

Orion Crew Survival System Suits

At several points during Artemis missions, astronauts will wear a bright orange spacesuit called the Orion Crew Survival System (OCSS) suit, which is designed to protect them on their journey.

Improvements have been made from head to toe to the suit worn on the space shuttle for Orion. Elements have been reengineered to improve safety and range of motion for astronauts, and instead of the small, medium, and large sizes from the shuttle era, they will be custom fit for each crew member.

The suits can keep astronauts alive for up to six days if Orion were to lose cabin pressure during its journey, with interfaces that supply air and remove carbon-dioxide. They are also equipped with a suite of survival gear in the event astronauts must exit Orion after splashdown in the ocean before recovery personnel arrive, and the color is easily recognizable in the ocean.

The outer layer is fire resistant and a stronger zipper allows astronauts to quickly put the suit on. Improved thermal management will help keep them cool and dry. A lighter, stronger helmet improves comfort and communication, and the gloves are more durable and touch-screen compatible. Better-fitting boots will also provide protection in the case of fire and help an astronaut move more swiftly.

Astronauts will wear the suit on launch day, in emergency situations, high-risk parts of missions near the Moon, and during the high-speed return to Earth. Its design and engineering enhancements provide an additional layer of protection for astronauts and ensure they return home safely from deep space missions.

▲ Dustin Gohmert, Orion Crew Survival Systems Project Manager at NASA's Johnson Space Center, poses for a portrait while wearing the Orion Crew Survival System (OCSS) suit on Oct. 15, 2019 at NASA Headquarters in Washington, DC.

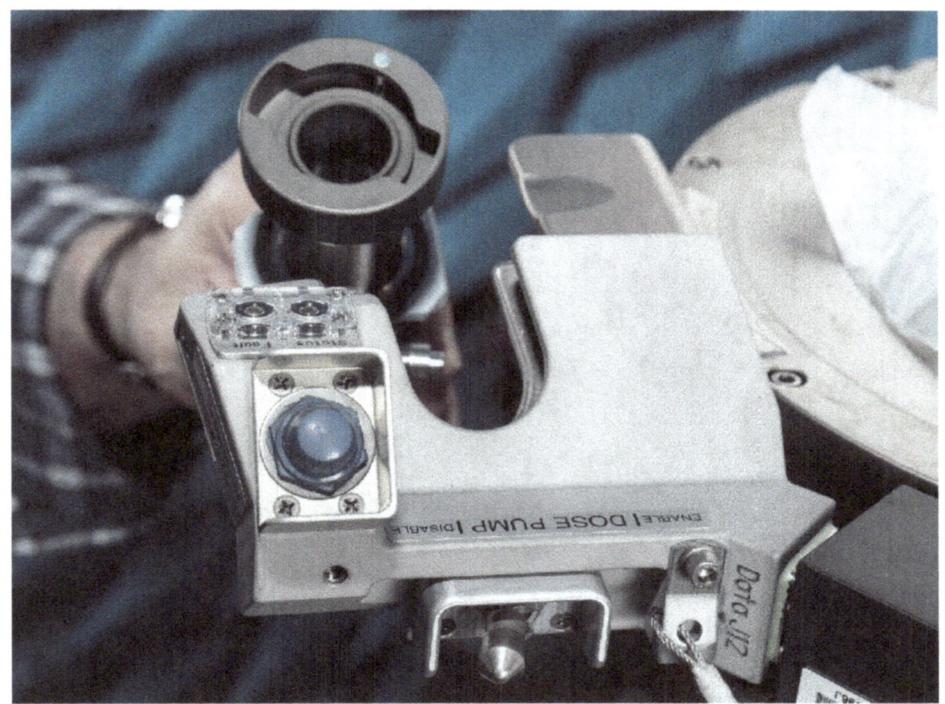

◄ A team member demonstrates using the Universal Waste Management System (UWMS), lifting the urine hose out of its cradled position like a crew member would for use. A funnel (not shown) is attached to the open end of this hose and can then be easily replaced or removed for disinfection.

Orion Crew Life Support and Safety Fun Facts

» Orion's closed loop life support system is capable of maintaining a positive pressure, breathable atmosphere, and thermal cooling for up to 144 hours to suited crewmembers in the event of a pressure vessel leak or contaminated cabin atmosphere.

» Orion is designed to execute an abort and return to Earth from any point in the mission regardless of orbital alignment with Earth, and provides a safe haven and survival equipment for onboard crew for up to 24 hours post-landing.

» Orion is designed to provide assured crew return in the event of multiple system failures incorporating a backup flight computer, manual control of life support systems, communication, and navigation aids.

» NASA and Lockheed Martin engineers working on the design of the Orion crew module collaborated with NASCAR and Indy Car racing experts specializing in crash injuries to design state-of-the art crew seats, restraints, and impact attenuation mechanisms to prevent injury of the astronauts in the event of a hard landing.

Apollo
to Artemis

While technology has improved since NASA's final Apollo mission almost 50 years ago, the underlying physics principles that dictated Apollo's shape and general design remain the same. That is largely why Apollo and Orion bear a striking visual resemblance from the outside. Building off what was learned from the Apollo program, Orion's major advancements will allow NASA to extend our reach past Earth to the Moon and beyond, expand the range of mission destinations, and accommodate larger crews for longer missions.

Orion and Apollo systems are similarly comprised of three major elements – a launch abort system, which protects the crew during launch and ascent; a crew module, which houses the crew and maintains a crew environment; and a service module, which provides main propulsive forces and supplies the crew module with power and life sustaining commodities.

Orion's launch abort system (LAS) looks like the escape tower flown on top of Apollo's Saturn V rocket, but upgrades make it possible for Orion to escape from its launcher under more extreme conditions. This is largely due to the attitude control motor on Orion's LAS, which consists of eight variable-thrust nozzles that can be independently controlled and modulated. In the event of an abort, Apollo's non-vectoring "pitch" motor would only send the capsule in one direction. Orion's computer, however, can sense the vehicle's orientation and send commands to the attitude control motor to adjust each nozzle's position and thrust, keeping Orion properly positioned away from the rocket.

At the system level, there are significant differences between Orion and Apollo. The Orion elements are larger in scale, offering greater volume to accommodate larger crews and provide greater range. Where Apollo was designed to transport three crew for 14 days, Orion can transport up to four crew for 21 days and is designed to be versatile and support a variety of destinations.

▲ On the left, the spacecraft for Apollo 10, with its Command and Service Modules attached, is moved to a work stand for mating to a lunar module adapter inside the Manned Spacecraft Operations Building (now the Neil Armstrong Operations and Checkout Building) at NASA's Kennedy Space Center in Florida on Jan. 31, 1969. To the right is the Orion spacecraft being lifted out of a processing work stand and onto a transporter inside the Neil Armstrong Operations and Checkout Building on Jan. 14, 2021.

On the left, NASA's 363-foot-tall Saturn V rocket for the Apollo 4 mission stands inside the Vehicle Assembly Building (VAB) at the agency's Kennedy Space Center in Florida in preparation for roll out to the launch pad. On the right, the 322-foot-tall Space Launch System rocket and Orion spacecraft for the Artemis I mission stands inside the VAB ahead of rollout for the mission's wet dress rehearsal.

Orion has 30 percent more habitable space than Apollo, allowing more room for the crew. Orion is also designed with many more crew comforts to support longer duration spaceflight, such as a galley for preparing meals and a functioning waste management system. Crew exercise was very difficult inside Apollo, as the environmental control system was not designed to compensate for crew workouts. However, Orion provides full exercise capability, and is also designed to maximize available privacy and mitigate noise and odors, enabling longer, healthier, and more hygienic missions. Modern features have also been incorporated into Orion, including composite materials, 3D printed parts, solar arrays, and an improved heat shield design.

Orion also has over 1,200 sensors, and the crew module has a glass cockpit with screens and user-interfaces reflective of our digital age, rather than Apollo's analog inputs and outputs.

Where Apollo ushered in the era of computers and software, Orion capitalizes on decades of computing advancements. Orion's guidance, navigation, and control (GN&C) system is comprised of flight computers, displays and controls, optic measurement, and advanced software. Compared to Apollo's single flight computer, Orion has two simultaneously operating redundant flight computers that each include two redundant computer modules, giving it a total of four redundant systems. In addition, just one of Orion's redundant computers is only 75 percent the weight of the sole computer aboard Apollo, has 128,000 times more memory and is 20,000 times faster. Orion's computing redundancies not only improve safety, but they also improve data collection and processing power.

Apollo's flight software was only capable of calculating the spacecraft's trajectory; it would not adjust it real-time in space based on sensor input, as Orion can. In contrast to the manual operation necessary on Apollo, Orion's software-driven avionics and other computers provide situational awareness and autonomy. Orion automates functions, thereby freeing astronauts for other tasks, as they will not have to frequently monitor spacecraft systems and verify their trajectory. The software also enables full uncrewed test flights and checkout before a mission is crewed, drastically increasing crew safety.

Relative to Apollo, Orion's systems have also been upgraded to account for the effects of space radiation exposure, an environmental hazard that is better understood since the Apollo missions. To counter the radiation hazard, electronic components within the computers are radiation hardened. This is something Apollo did not account for as missions were flown during a period of minimum solar activity.

The power system on Orion enables an entirely new class of missions when compared to Apollo. Orion provides a renewable power supply using solar cells to capture energy from the Sun, where Apollo generated power using a finite supply of hydrogen and oxygen loaded at the start of the mission to harness energy as they combined in a fuel cell. Free from the constraint of a resource-limited power supply, Orion pioneers more distant crewed exploration and sustains extended mission durations.

Orion's parachutes may also look like those used during the Apollo-era, but through testing and analysis, technicians have developed Orion's parachutes to be lighter, their performance better understood, and more capable than Apollo's. Engineers have figured out how to manage the stresses on the system during deployment more efficiently, decrease the mass of the parachutes by using high tech fabric materials rather than metal cables for the risers that attach the parachute to the spacecraft, and improve how the parachute is packed into Orion so they deploy more reliably.

There are a number of differences between Orion and Apollo's crew module uprighting system (CMUS), the airbags responsible for uprighting the crew module in the case of an inverted splashdown. Due to Orion's larger diameter as compared to Apollo, five airbags are required to upright Orion, whereas the smaller Apollo only required three airbags. The Apollo airbags were inflated using a compressor and solenoid valves, while Orion's CMUS uses pyro-valves and helium to inflate the airbags. As a result, the Orion CMUS inflation system is lighter than the Apollo system.

Space Launch System

NASA's Space Launch System, or SLS, is an evolvable family of super heavy lift launch vehicles that provides the foundation for human exploration beyond Earth's orbit. With its unprecedented power and volume for payloads, SLS is the only rocket that can send Orion, astronauts, and cargo to the Moon on a single mission.

Offering more payload mass, volume, and departure energy per launch than commercial rockets, SLS is designed to be flexible and evolvable and will open new possibilities for payloads, including robotic scientific missions to places like the Moon, Mars, the gas giants Saturn and Jupiter, and the ice giants Uranus and Neptune.

The SLS team is producing NASA's first deep space rocket built for human space travel since the Saturn V. Each variant in the evolvable SLS family uses the same main propulsion system of a core stage, liquid fueled engines, and solid rocket boosters. The first SLS vehicle, called Block 1, will create more than 8.8 million pounds of thrust during launch and ascent and will be capable of delivering more than 27 metric tons (59,525 pounds) to the Moon. It will be powered by twin five-segment solid rocket boosters and four RS-25 liquid propellant core stage engines. After reaching space, the interim cryogenic propulsion stage (ICPS) sends Orion on to the Moon. The first three Artemis missions will use a Block 1 rocket with an ICPS.

The next planned evolution of the SLS, the Block 1B crew vehicle, will replace the ICPS with a new, more powerful exploration upper stage (EUS) to enable more ambitious missions. The Block 1B vehicle can send 38 tons (nearly 84,000 pounds) to the Moon and in a single launch carry the Orion spacecraft along with large cargos for exploration systems needed to support a sustained presence on the Moon.

Boeing is the prime contractor for the core stage, ICPS, and EUS; Aerojet Rocketdyne built the RS-25 engines for those stages. Northrop Grumman is the prime contractor for the five-segment solid rocket boosters, which are the largest ever built for spaceflight. Teledyne Brown Engineering is the lead contractor for the launch vehicle stage adapter (LVSA), and Marshall Space Flight Center built the Orion stage adapter (OSA).

Solid Rocket Boosters

Together, the SLS twin solid rocket boosters provide more than 75 percent of the total SLS thrust at launch. The SLS booster design is based on the space shuttle boosters but has an additional motor propellant segment to provide more power and several other upgrades. The new design has completed five full-scale test firings in a horizontal position with the final qualification motor test in June 2016. Standing 177 feet tall, the SLS booster is the largest, most powerful solid propellant booster ever built for flight, with more than three million pounds of thrust each. While the boosters are using metal casings and parts that were flown on space shuttle missions, they have many new upgrades and new parts, including a new insulation-liner configuration and new avionics systems to control flight. In addition, the SLS boosters undergo an improved nondestructive evaluation process to confirm each motor's readiness for launch.

NASA and booster prime contractor Northrop Grumman are working to design, develop and test next-generation boosters that will power SLS flights after all available shuttle-era hardware is expended. Teams are testing small- scale solid rocket motors to evaluate new components including propellant, insulation materials, and nozzle materials for future SLS Block 2 flights.

RS-25 Engine

Four Aerojet Rocketdyne RS-25 engines will power the core stage of the SLS. The liquid hydrogen/liquid oxygen- powered engines are compact and have high performance. At full throttle, the four engines will give the SLS about two million of its 8.8 million pounds of maximum thrust

▲   Inside the Rotation, Processing and Surge Facility at NASA's Kennedy Space Center in Florida on Nov. 17, 2020, the left and right booster segments for the Space Launch System are being prepared for their move to the Vehicle Assembly Building (VAB).

for the full eight-minute climb to orbit. Although SLS uses RS-25 engines that flew on the space shuttle, they had to be modified slightly and tested to fly on the SLS rocket. The acceptance testing is complete for all 16 RS-25 engines that served as former space shuttle main engines. Tests at NASA's Stennis Space Center near Bay St. Louis, Mississippi have shown that the engines can perform at the higher power level and other unique requirements needed to launch the super heavy-lift SLS rocket. Since January 2015, NASA has conducted 32 developmental and flight engine tests for a total of more than 14,000 seconds – more than four hours – of cumulative hot fire. Testing included certifying new engine controllers to be used by the heritage RS-25 engines, Green Run testing of two new, unflown engines, and tests with new parts for future engines. Future-production engines will include parts made with advanced manufacturing techniques, such as additive manufacturing, that have the potential to streamline cost and production of the engines.

Core Stage

The SLS core stage is the world's tallest rocket stage. Towering 212 feet with a diameter of 27.6 feet, it will store cryogenic liquid hydrogen and liquid oxygen and all the systems that will feed the stage's four RS-25 engines. It also houses the flight computers and other avionics needed to control the rocket's flight. The core stage is designed to operate for approximately 500 seconds, reaching nearly Mach 23 and more than 530,000 feet in altitude before it separates from the upper stage and Orion spacecraft. The core stage serves as the backbone of the rocket, supporting the weight of the payload, upper stage, and Orion, as well as the thrust of its four RS-25 engines and two five-segment solid rocket boosters attached to the engine and intertank sections. The core stage was built by Boeing at NASA's Michoud Assembly Facility near New Orleans using the world's largest spacecraft welding tool, the Vertical Assembly Center among other advanced spacecraft manufacturing and welding tools.

Core Stage Green Run

The Artemis I core stage with its four RS-25 engines underwent a series of "green run" tests in 2020 and early 2021 to thoroughly checkout this brand-new rocket stage that has never flown before. After manufacturing at NASA's Michoud Assembly Facility near New Orleans, Louisiana, the stage was barged to NASA's Stennis Space Center in Mississippi and installed in the B-2 test stand. There were eight major test cases that made up the Green Run series. The first, a modal test that applied forces to the stage to validate structural loads computer models was completed on Jan. 30, 2020.

The second test was a checkout of the avionics and flight computers, which control the rocket's first eight minutes of flight. The third test exercised all the safety systems that shut down the stage in case of an issue. The fourth test of the series tested each of the main propulsion system components that connect to the engines. During the fifth test, engineers ensured that the thrust vector control system steers the four engines and checked all the related hydraulic systems. The sixth test simulated the launch countdown, including step-by-step propellant loading procedures, and the test team exercised and validated the countdown timeline and sequence of events. During the seventh test, known as the wet dress rehearsal, engineers demonstrated loading, controlling, and draining more than 700,000 gallons of cryogenic propellants into and out of the core stage and then returning the stage to a safe condition.

The Green Run testing culminated with all four engines being fired in early 2021 for eight minutes, just as they will operate during launch to send Orion and the upper stage to space.

Following the test, NASA shipped the core stage to the agency's Kennedy Space Center in Florida for final assembly and integration with the Orion spacecraft.

Interim Cryogenic Propulsion Stage

The interim cryogenic propulsion stage (ICPS) is a 5-meter, single-engine liquid hydrogen/ liquid oxygen- based system that provides in-space propulsion after the solid rocket boosters and core stage are jettisoned. The ICPS, built by Boeing and United Launch Alliance, is a modified Delta Cryogenic Second Stage, a proven upper stage used on United Launch Alliance's Delta IV family of launch vehicles. Modifications for SLS included lengthening the liquid hydrogen tank, adding hydrazine bottles for attitude control, and minor avionics changes. The ICPS is powered by one Aerojet Rocketdyne RL10 engine and generates 24,750 pounds of maximum thrust.

Launch Vehicle Stage Adapter

The cone-shaped launch vehicle stage adapter (LVSA) partially encloses the ICPS and connects it to the SLS core stage below and the Orion stage adapter (OSA) above. A pneumatically actuated frangible joint assembly at the top of the adapter separates the core stage and LVSA from the ICPS, OSA, and Orion. In addition to providing structural support for launch and a separation system, the LVSA also protects avionics and electrical devices in the ICPS from extreme vibration and acoustic conditions during launch and ascent. Prime contractor Teledyne Brown Engineering manufactures the LVSA using self- reacting friction-stir welding tools at NASA's Marshall Space Flight Center in Huntsville, Alabama.

Orion Stage Adapter

The Orion stage adapter (OSA), built in-house by Marshall Space Flight Center, connects the ICPS to the Orion spacecraft. The adapter contains a diaphragm that provides a barrier to prevent gases generated during launch, such as hydrogen, from entering Orion. The OSA can also carry small payloads, called CubeSats, to deep space. The OSA can potentially accommodate up to 17 CubeSats in a combination of 6U and 12U sizes (one unit, or U, is 10 cm x 10 cm x 10 cm). The actual number of CubeSats manifested on a flight depends on several factors, including mission parameters and the combined weight of these small spacecraft. The SLS Program provides a comprehensive secondary payload deployment system for CubeSats, including mounting brackets for commercial off the-shelf dispensers, cable harnesses, a vibration isolation system, and an avionics unit. CubeSats can play a key role in the Artemis missions by gathering data and demonstrating potential technologies that reduce risk, increase effectiveness, and improve the design of robotic and human space exploration missions.

Structural Testing

Structural testing at NASA's Marshall Space Flight Center in Huntsville, Alabama, has been completed on the 212- foot SLS core stage and the upper part of the rocket, which includes the stage that sends the Orion spacecraft to the Moon. These tests help verify computer models showing the structural design can survive flight. Special test stands subjected the core stage engine section, liquid hydrogen tank, intertank, and liquid oxygen tank to millions of pounds of flight-like pushing, pulling, bending and twisting stresses. The structural test program, which was the most extensive structural test program since the space shuttle, began in 2017 with the engine section and ended in 2020 with the liquid oxygen tank.

In a separate test program completed in 2017, the SLS team performed similar structural testing for the SLS in-space stage and payload sections, including the ICPS, the LVSA and the OSA, as well as a frangible joint assembly. The adapters connect the parts of the vehicle, change its diameter, and protect electronics during ascent. The ICPS, powered by one Aerojet Rocketdyne RL10 engine, helps send Orion on its trajectory to the Moon. A previous version of the RL10 and the Delta Cryogenic Second Stage, on which the ICPS is based, have both served on numerous flights of the United Launch Alliance Delta IV rocket with the RL10 just completing its 500th flight; both have completed extensive testing.

Flight Software and Avionics

The core stage flight computers and avionics, which control the rocket's first eight minutes of flight, have undergone extensive qualification testing both in lab testing at NASA's Marshall Space Flight Center, on the Artemis I core stage at Michoud Assembly Facility and during Green Run stage testing at Stennis Space Center. In the lab environment, the flight software has been tested by flying thousands of simulated launches and flights replicating both nominal and contingency operations. Avionics testing has been completed for the core stage and for the other SLS propulsion element that launches the rocket, the twin solid rocket boosters.

Wind Tunnel Testing

The SLS team completed numerous tests that simulate launch and flight conditions. Extensive wind tunnel testing has been conducted at NASA's Ames Research Center in Moffett Field California, Langley Research Center in Hampton Virginia, and Marshall Space Flight Center in Huntsville, Alabama. These tests used scale rocket models to study how the vibration, base heating, and other environments affect the launch vehicle designs for both the Block 1 and Block 1B SLS configurations. Scale models of the upgraded rocket in crew and cargo configurations were carefully positioned in wind tunnels for test programs to obtain data needed to refine the design of the rocket and its guidance and control systems. During hundreds of test runs at Langley and Ames, engineers worked on measuring the forces and loads that air induces on the launch vehicle during every phase of its mission.

Acoustic Testing

Researchers at NASA's Neil A. Armstrong Test Facility, formerly known as Plum Brook Station, in Sandusky, Ohio, completed a development test on a proposed design of acoustic panels for the SLS universal stage adapter, an adapter that will connect an exploration upper stage (EUS) planned for future SLS configurations with the Orion spacecraft. Given the extreme sound produced by the world's most powerful rocket, this test series, conducted at Armstrong's Reverberant Acoustic Test Facility, provided data for acoustic modeling that will be used to ensure future payloads aboard the Block 1B configuration are protected from the high levels of noise and vibration experienced during launch. Even the Block 1 configuration of the rocket generates significant acoustic vibrations. Acoustic testing with a scale model of the SLS Block 1 was conducted early in the program at NASA's Marshall Space Flight Center in Huntsville, Alabama.

SLS Fun Facts

» At liftoff, the Block 1 Space Launch System (SLS) will provide 8.8 million pounds of thrust.

    » This is more than 32 times the total thrust of a 747 jet.

    » It's also 15 percent more thrust than the Saturn V at liftoff.

» 8.8 million pounds of thrust produces horsepower equivalent to:

    » 167,600 Corvette engines

    » 14,000 locomotive engines

» The Block 1 configuration will stand 322 feet tall, which is taller than the Statue of Liberty.

» At launch, the SLS will weigh 5.75 million pounds, which is equivalent to more than 7.5 fully loaded 747 jets.

» The SLS can bring 59,525 pounds to the Moon. This is equivalent to nearly 30 one-ton pickup trucks' worth of cargo, or 4.5 fully grown elephants.

» If the heat energy of the SLS solid rocket boosters (SRBs) could be converted to electric power, the two SRBs firing for 2 minutes would produce 2.3 million kilowatt hours of power, enough to supply power to over 92,000 homes for a full day.

» Each SRB burns approximately 5.5 tons of propellant per second.

SLS Quick Facts

Solid Rocket Booster

Length	177 ft.
Diameter	12 ft.
Weight	1.6 million lbs. each
Propellant	Polybutadiene acrylonitrile (PBAN)
Thrust	3.6 million lbs. each
Operational Time	126 seconds

Core Stage

Length	Approximately 212 ft.
Diameter	27.6 ft.
Empty Weight	Approximately 188,000 lbs.
Capacities	Liquid hydrogen (LH2) 537,000 gallons (2 million liters) (317,000 lbs.) Liquid oxygen (LOX) 196,000 gallons (742,000 liters) (1.86 million lbs.)
Material	Aluminum 2219

Interim Cryogenic Propulsion Stage (ICPS)

Height	45 ft.
Diameter	17 ft.
Propellant	Liquid hydrogen/liquid oxygen
Engine	1 Aerojet Rocketdyne RL10
Thrust	24,750 lbs.

RS-25 Engine

Thrust	177 ft.
Size	12 ft.
Weight	1.6 million lbs. each
Operational Time	Polybutadiene acrylonitrile (PBAN)
Operational Temp Range	3.6 million lbs. each

Orion Stage Adapter (OSA)

Height	5 ft.
Top Diameter	18 ft.
Payload Volume	516 ft³ up to 17 berths for 6U/12U

Launch Vehicle Stage Adapter (LVSA)

Height	27.5 ft.
Top Diameter	16.5 ft.
Bottom Diameter	27.5 ft.

▲   In High Bay 3 of the Vehicle Assembly Building at NASA's Kennedy Space Center in Florida, the right-hand forward segment is lowered by crane on top   of the center forward segment on the mobile launcher for the Space Launch System (SLS) on March 3, 2021.

Exploration Ground Systems

The Exploration Ground Systems Program (EGS) is one of four NASA programs based at the agency's Kennedy Space Center in Florida, including Launch Services, Commercial Crew, and Exploration Research Technology programs. EGS was established to develop and operate the systems and facilities necessary to process and launch rockets and spacecraft during assembly, transport, launch, and recovery.

EGS's mission is to transform the center from a historically government-only launch complex to a spaceport that can handle several different kinds of spacecraft and rockets and NASA's exploration objectives by developing the necessary ground systems, infrastructure and operational approaches.

Unlike previous work focusing on a single kind of launch vehicle, such as the Saturn V or space shuttle, engineers and managers in EGS are preparing infrastructure to support several different kinds of spacecraft and rockets that are in development, including NASA's Space Launch System (SLS) rocket and Orion spacecraft for the Artemis program.

A key aspect of the program's approach to long-term sustainability and affordability is to make processing and launch infrastructure available to commercial and other government customers, thereby distributing the cost among multiple users and reducing the cost of access to space.

NASA is developing the SLS exploration class rocket and working with several private companies to produce vehicles to take astronauts to low-Earth orbit and the International

In the transfer aisle of the Vehicle Assembly Building at NASA's Kennedy Space Center in Florida, workers use a crane to lift up the left-hand booster forward assembly for the agency's Space Launch System for transfer into High Bay 3 on March 1, 2021.

Multi-Element Verification and Validation: Mobile Launcher/ Launch Pad 39B

The mobile launcher — the 380-foot-tall ground structure that will be used to assemble, process, and launch SLS — has gone through a series of tests both in the Vehicle Assembly Building (VAB), the facility where space vehicle components are assembled and stacked vertically onto the mobile launcher platforms, and the newly renovated Launch Pad 39B, the pad from which Artemis missions will be launched. Crawler-Transporter 2 (CT-2) completed a test rollout of the mobile launcher for integrated testing at the newly renovated Launch Pad 39B, validating it can communicate effectively with the facility systems and ground systems to perform appropriately during launch. CT-2 rolled the mobile launcher to Launch Pad 39B in June 2019, where it stayed for several months for multi- element verification and validation tests. This included the sound suppression system, loading the cryogenic fuel system, a simultaneous umbilical swing test, and more. CT-2 returned the mobile launcher to the VAB in December 2019 for modal testing with the mobile launcher sitting on the mount pedestals as well as CT-2 sharing the load. Exploration Ground Systems conducted a six-hour pressurization test of the liquid oxygen (LOX) tank at Launch Pad 39B, which has been upgraded for the agency's SLS rocket. The SLS will use both liquid oxygen and liquid hydrogen propellants. An initial test of Xenon lights took place with the mobile launcher on the pad.

Space Station. The SLS is set to be the most powerful U.S. rocket since the Saturn V took astronauts to the Moon and will act as the cornerstone for NASA's future human exploration missions to deep space destinations.

EGS is focusing on the equipment, management and operations required to safely connect a spacecraft with a rocket, move the launch vehicle to the launch pad, successfully launch it into space, and recover the spacecraft. The work entails the use of many of the facilities unique to Kennedy, such as the 52-story Vehicle Assembly Building and Launch Pad 39B.

Kennedy has more than 50 years serving as our nation's gateway to exploring the universe. Taking the knowledge and assets of NASA's successful spacefaring past, the EGS Program is helping to build a successful and diverse future in spaceflight.

The EGS and Jacobs teams rolled the mobile launcher atop CT-2 out of the VAB to Launch Pad 39B for a second time in October 2020, to help prepare the launch team for the "wet dress rehearsal," when SLS and Orion are rolled out to the pad atop the mobile launcher to practice fueling operations shortly before launch. During the mobile launcher's 10-day stay, engineers performed several tasks, including a timing test to validate the launch team's countdown timeline, and a thorough, top- to-bottom wash down of the mobile launcher to remove any debris remaining from construction and installation of the umbilical arms.

Launch Control Center
Young-Crippen Firing Room 1 / Launch Simulations

The firing rooms are the heart of the Spaceport Command and Control System at Kennedy, which will operate, monitor, and coordinate the ground equipment for launch of the SLS rocket and Orion spacecraft. All the activities involved with preparing rockets, spacecraft, and payloads for space can be controlled by engineers sitting at computer terminals in the firing rooms. Likewise, all activities at the launch pads can be run from a firing room.

The Young-Crippen Firing Room, previously known as Firing Room 1, oversaw launches ranging from the first Apollo missions to the first space shuttle mission and, after many upgrades, will be used for Artemis I.

The Artemis I launch team has conducted a series of simulations to test critical portions of the countdown to ensure everyone is ready to handle any situation launch

day throws their way. Under the leadership of Launch Director Charlie Blackwell-Thompson, the team has tested cryogenic loading, terminal count, otherwise known as the final countdown to launch, and other portions of the launch software currently in the final phases of development.

Spacecraft Recovery

Exploration Ground Systems leads a joint NASA and Department of Defense team that continues training at sea to improve joint landing and recovery operations for the Orion spacecraft and crew from future deep space exploration missions. Recovery personnel have practiced procedures and operations in the Neutral Buoyancy Laboratory pool near Johnson Space Center, and in the open water off the coast of California, during a series of Underway Recovery Tests using a test version of the Orion spacecraft and other equipment that will be used during recovery operations.

▼ Members of the Artemis I launch team participate in a countdown simulation inside the Launch Control Center's Firing Room 1 at NASA's Kennedy Space Center in Florida on Feb. 3, 2020.

▲   Inside the Multi-Operations Support Building near the Multi-Payload Processing Facility (MPPF) at NASA's Kennedy Space Center in Florida on Aug. 16, 2019, a row of Self-Contained Atmospheric Protective Ensemble (SCAPE) suits hang inside a changing room.

These tests help to evaluate and improve recovery procedures and hardware ahead of Orion's flight on Artemis I and II. Testing also included recovering the spacecraft from Orion's first flight test, Exploration Flight Test-1, when the spacecraft launched atop a Delta IV Heavy rocket on Dec. 5, 2014, from Space Launch Complex 37 at Cape Canaveral Air Force Station in Florida before splashing down in the Pacific Ocean about 600 miles off the coast of San Diego. The Landing and Recovery team completed their final certification run in 2021 to ensure they are ready for splashdown of Orion after the Artemis I mission.

SCAPE

Self-Contained Atmospheric Protective Ensemble (SCAPE) suits are used in operations involving toxic propellants and are supplied with air either through a hardline or through a self-contained environmental control unit. During preparations for launch, teams at Kennedy Space Center are responsible for loading the Orion vehicle with propellants prior to transportation to the Vehicle Assembly Building, where it will be secured atop the Space Launch System rocket.

Exploration Ground Systems spent time preparing for Artemis I with a series of hazardous hypergolic test events at the Kennedy Multi-Payload Processing Facility (MPPF) prior to the spacecraft's arrival. In August 2019, SCAPE technicians practiced by putting on the 38 suits for a test simulation of loading propellants into a replicated test tank for Orion. After donning their suits, the technicians completed a tanking to test the system before Orion arrived for processing.

ULA

STORRM / STS-134

The Sensor Test for Orion Relative Navigation Risk Mitigation (STORRM) was successfully demonstrated on Space Shuttle Endeavour's STS-134 mission to the International Space Station (ISS) on May 16, 2011.

The goal of STORRM was to validate a new relative navigation sensor based on advanced laser and detector technology that will make docking and undocking easier and safer, and test the hardware in the same environment that the sensors would experience on the first Orion rendezvous to another vehicle. On Orion, these sensors were installed on the hatch window of the crew module.

Astronauts tested the new navigation and docking system on two different flight days, first on their initial approach to dock to the ISS and later after undocking from the ISS. STS-134 performed an unprecedented re-rendezvous of the ISS, as Commander Mark Kelly and pilot Greg Johnson navigated the shuttle to make an Orion-like approach to the ISS before de-orbiting on Flight Day 14. Astronaut Andrew Feustel operated the STORRM equipment during the testing.

STORRM hardware included a laser-based state-of-the art vision navigation sensor (VNS), which provided an image of the target (in this case, the ISS) and calculated with precise accuracies for range, bearing, alignment and orientation data up to six degrees of freedom. In comparison, the space shuttle's sensors could only calculate up to three degrees of freedom. The STORRM hardware also included a docking camera to provide high-resolution color imagery.

Working together, these sensors provided real-time 3D images with a resolution 16 times higher than previous space shuttle sensors. An avionics assembly, which consisted of a power distribution unit, data recorder unit, and memory storage, was manifested to communicate with a space-certified laptop. Five reflective docking targets were previously installed on the space station during STS-131 and provided a cooperative target for the VNS.

STORRM collected 600 gigabytes of data and the VNS performed better than expected, providing continuous measurements from as far away as 3.5 miles to within six feet of the space station – three times the range capability of the current relative navigation sensor. The sensor technology also provided exceptional 3D images of the target.

Teams conduct powerup and docking operations for the Sensor Test for Orion Relative Navigation Risk Mitigation (STORRM) in a payload support room at Johnson Space Center's Mission Control Center in Houston on May 18, 2011.

The data collected by STORRM will help to make docking and un-docking with other spacecraft safer and easier for astronauts on Orion.

STORRM was developed by the Orion Program at NASA's Johnson Space Center, which was responsible for program management, technology evaluation, flight test objectives, operational concepts, contract management, and data post-processing.

Engineers at NASA's Langley Research Center were responsible for engineering management, design and build of the avionics assembly, STORRM software application, and reflective elements. They were also responsible for the integration, testing, and certification of these components. Industry partners Lockheed Martin and Ball Aerospace were responsible for the design, build, and testing of the VNS and docking camera.

Launch Date	May 16, 2011 8:56 a.m. EDT
Landing Date	June 1, 2011 2:34 a.m. EST
Launch Site	Kennedy Space Center Launch Complex 39A
Launch Vehicle	Space Shuttle Endeavour
Landing Site	Kennedy Space Center

Pad Abort-1

On May 6, 2010, at the U.S. Army's White Sands Missile Range (WSMR) near Las Cruces, New Mexico, NASA conducted a successful flight test known as Pad Abort-1, a test of the Orion launch abort system (LAS), a critical safety system that sits atop Orion at launch and during ascent. If an emergency occurs during launch, the LAS pulls Orion and its crew away from the rocket for a safe landing in the Atlantic Ocean. Pad Abort-1 demonstrated the capability of the LAS to propel the crew module to a safe distance during a ground-initiated abort on the launch pad. The flight test lasted about 135 seconds from launch until the crew module touched down about a mile north of the launch pad.

The test involved three motors. An abort motor produced a momentary half-million pounds of thrust to propel the crew module away from the pad. It burned for approximately six seconds, with the highest impulse in the first 2.5 seconds. The crew module reached a speed of approximately 445 mph in the first three seconds, with a maximum velocity of 539 mph in its upward trajectory to about 1.2 miles high.

The attitude control motor fired simultaneously with the abort motor and steered the vehicle, using eight thrusters that produce up to 7,000 pounds of thrust. It provided adjustable thrust to keep the crew module on a controlled flight path and reoriented the vehicle as the abort system burned out.

The jettison motor, the only motor of the three that would be used in all nominal rocket launches, pulled the entire launch abort system away from the crew module and cleared the way for parachute deployment and landing. After explosive bolts fired and the jettison motor separated the system from the crew module, the recovery parachute

▼ This aerial scene following the Pad Abort-1 test on May 6, 2010 shows the crew module and main parachutes on the ground at the White Sands Missile Range in New Mexico.

Orion Abort Scenario

system deployed. The parachutes guided the crew module to touchdown at 16.2 mph (24 feet per second), about one mile from the launch pad.

The Orion Program at NASA's Johnson Space Center in Houston led the launch abort system test team. NASA's Langley Research Center designed and built the boilerplate crew module for the test, and NASA's Dryden Flight Research Center prepared the crew module for integration and led the flight test vehicle integration at WSMR with Lockheed Martin, the prime contractor to NASA for Orion.

Launch Date	May 6, 2010, 7:00 a.m. MDT
Launch Site	U.S. Army's White Sands Missile Range, New Mexico Launch Complex 32E
Duration	2 min., 15 sec.
Distance	6,900 ft. downrange
Altitude	6,336 ft.

Exploration Flight Test-1

On Dec. 5, 2014, Orion lifted off from Space Launch Complex 37 at Cape Canaveral Air Force Station in Florida on a United Launch Alliance Delta IV Heavy rocket and splashed down approximately 4.5 hours later in the Pacific Ocean. The test, called Exploration Flight Test-1, or EFT-1, demonstrated Orion's space-worthiness, tested the spacecraft's heat shield during entry into Earth's atmosphere, and proved the capsule's recovery systems. The test flight carried a high-fidelity crew module and a mock service module that remained attached to the rocket's upper stage until its return to Earth.

At 7:05 a.m., the three-core first stage of the Delta-IV Heavy rocket ignited, lifting Orion off to begin the mission. Approximately four minutes after liftoff, the two side boosters separated as the center core continued firing. The second stage ignited after separation to begin the first of three planned burns. During the first burn, the service module's protective fairing separated, followed by the launch abort system. This first burn of the second stage placed the spacecraft into a preliminary 115-by-552-mile

parking orbit. After completing one revolution around the Earth, during which time controllers in Mission Control in Houston verified the functioning of the spacecraft's systems, the second stage ignited a second time, raising Orion's apogee, or high point, above the Earth to 3,600 miles. During the coast to apogee, Orion remained attached to the second stage and completed its first crossing through the inner Van Allen radiation belt.

A Delta IV Heavy rocket lifts off from Space Launch Complex 37 at Cape Canaveral Air Force Station in Florida carrying NASA's Orion spacecraft on Exploration Flight Test-1, an unpiloted flight test to Earth orbit.

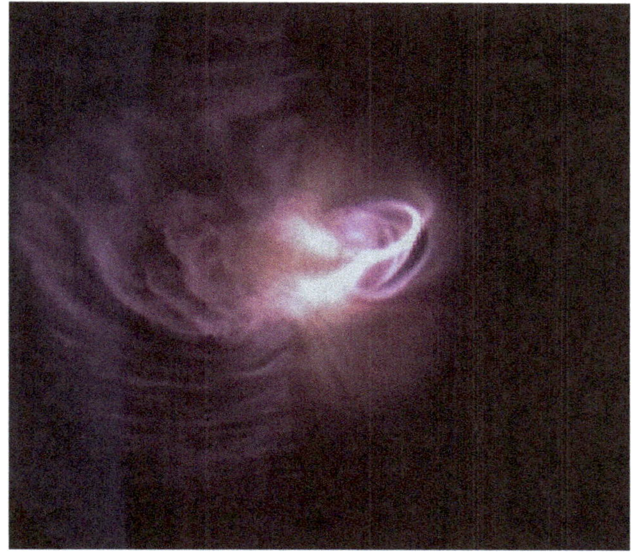

◀ A view from the top hatch window as plasma surrounds the vehicle during reentry on Orion's first flight test, Exploration Flight Test-1 (EFT-1), on December 5, 2014.

About three hours after launch, Orion reached its apogee and began its descent back toward Earth, separating from the second stage. During the passage back through the Van Allen belt, Orion fired its thrusters to adjust its course for return, and at an altitude of 400,000 feet, the spacecraft entered the Earth's atmosphere, traveling at 20,000 mph. The spacecraft experienced maximum heating of about 4,000 degrees Fahrenheit, proving the worthiness of the heat shield. After release of Orion's forward bay cover, two drogue parachutes deployed to slow and stabilize the spacecraft. Then the three main parachutes deployed, slowing the spacecraft to 20 mph. Splashdown occurred 4 hours and 24 minutes after launch about 600 miles southwest of San Diego.

A NASA and U.S. Navy team aboard the USS Anchorage recovered Orion and returned it to U.S. Naval Base San Diego. After its arrival, workers placed the Orion capsule aboard a truck that delivered it to the Kennedy Space Center. After engineers conducted a thorough

inspection of the spacecraft at Kennedy, workers trucked it to the Lockheed Martin facility in Littleton, Colorado, where engineers completed final inspections and decontamination of the vehicle, which is now on display at the Kennedy Space Center Visitors Complex.

EFT-1 collected data that is helping NASA lower risks to the astronauts who will later fly on Orion. This included verifying the thermal protection system, hardware separation events, the parachute system and the crew module uprighting system, as well as allowing engineers to identify design issues and fix them before Orion carries astronauts on future missions. The first flight also gave NASA the chance to continue refining its production and coordination processes, and Orion's design teams gained important experience and training to ensure the industry is prepared to launch Orion on its first integrated flight with the Space Launch System rocket. Lockheed Martin, NASA's prime contractor for Orion, manufactured the high-fidelity crew module and a mock service module. The Delta IV Heavy rocket was built by United Launch Alliance. Lockheed Martin began manufacturing the crew module in 2011, and it was constructed at NASA's Michoud Assembly Facility. Lockheed Martin delivered the crew module in July 2012 to the Neil Armstrong Operations & Checkout Building (O&C) at Kennedy Space Center, where heat shield installation, final assembly, integration, and testing were completed. More than 1,000 companies across the country manufactured or contributed elements to Orion for EFT-1.

Launch Date	Dec. 5, 2014, 7:05 a.m. EST
Landing Date	Dec. 5, 2014, 11:29 a.m. EST
Launch Site	Cape Canaveral Air Force Station, Florida, SLC-37B
Launch Vehicle	Delta IV Heavy
Landing Site	Pacific Ocean, 640 miles SSE off coast of San Diego; Recovered by USS Anchorage
Duration	4 hr., 24 min., 46 sec.

Ascent Abort-2

On July 2, 2019, NASA conducted a successful test known as Ascent Abort-2 (AA-2), which tested the Orion launch abort system (LAS), a critical safety system that sits atop Orion at launch and during ascent. If an emergency occurs during launch, the LAS pulls Orion and its crew away from the rocket for a safe landing in the Atlantic Ocean. Ascent Abort-2 demonstrated this capability of the LAS, propelling the crew module to a safe distance during ascent at a point midway between maximum aerodynamic pressure and maximum aerodynamic drag. This stressing test case, combined with subsystem qualification tests and the successful Pad Abort-1 flight test in 2010, resulted in a LAS that is certified to fly on the Artemis missions with astronauts on board.

During AA-2, a test booster provided by Northrop Grumman launched from Space Launch Complex 46 at Cape Canaveral Air Force Station in Florida at 7 a.m. EDT, carrying a fully functional LAS and a 22,000- pound Orion test vehicle to an altitude of 31,000 feet (approximately six miles) at Mach 1.18 (over 1,000 mph). At that point, the LAS' powerful reverse-flow abort motor fired 400,000 pounds of thrust, propelling the Orion test vehicle to a safe distance away from the rocket to splashdown in the Atlantic Ocean. The test took approximately three minutes.

A total of 890 developmental flight instrumentation (DFI) measurements were monitored and recorded during AA-2 on the booster, separation ring, crew module, and LAS. During descent, data from the test was downlinked as well as recorded on board by 12 data recorders that were ejected from the crew module in pairs starting 20 seconds after LAS jettison. The foam-encased data recorders initiated tracking beacons upon impact and floated in the ocean, enabling their recovery post-test.

Since AA-2 was meant to test Orion's ability to abort during ascent, it was not equipped with parachutes nor was the test capsule recovered from the ocean for cost-saving purposes. NASA already fully qualified the parachute system for flights with crew through an extensive series of 17 developmental tests and 8 qualification tests completed at the end of 2018.

▲ A fully functional launch abort system (LAS) with a test version of Orion attached, launches on NASA's Ascent Abort-2 (AA-2) atop a Northrop Grumman provided booster on July 2, 2019, at 7
a.m. EDT from Launch Pad 46 at Cape Canaveral Air Force Station in Florida.

NASA's prime contractor, Lockheed Martin, provided the fully functional Orion LAS, and the crew module to service module umbilical and flight design retention and release mechanisms. NASA's Johnson Space Center was responsible for producing the fully assembled and integrated crew module and separation ring. This included development of unique avionics, power, guidance, navigation and control (GN&C), software and data collection subsystems, and the ejectable data recorders. Johnson also provided the operations team responsible

for conducting the launch, as well as several elements of spacecraft on the launch pad.

The agency's Langley Research Center built the primary structure of the crew module test article and a separation ring that connected the test capsule to the booster and provided space and volume for separation mechanisms and instrumentation. Critical sensors and instruments used to gather data during the test were provided by NASA's Armstrong Flight Research Center. Exploration Ground Systems (EGS) and Jacobs Test and Operations Contract (TOSC) processed the vehicle at NASA's Kennedy Space Center before launch.

Launch Date	July 2, 2019, 7:00 a.m. EDT
Launch Site	Cape Canaveral Air Force Station, Florida, SLC-46
Duration	3 min., 13 sec.
Altitude	31,000 ft. (at abort initiation), 55,000 ft. (Orion test vehicle maximum altitude)

▼ The Ascent Abort-2 test vehicle is secured on the pad at Launch Complex 46 at Cape Canaveral Air Force Station in Florida after rollback of the Vertical Integration Facility on July 1, 2019.

Artemis I

Artemis I will be the first integrated flight of NASA's new deep space human exploration system — the Orion spacecraft, the Space Launch System (SLS) rocket, and Exploration Ground Systems at Kennedy Space Center in Florida — and will test our capabilities to orbit the Moon and return to Earth and set the stage for future missions to the lunar vicinity.

During this flight, an uncrewed Orion will launch on the most powerful rocket NASA has ever developed and fly farther than any spacecraft designed and built for humans has ever flown. On an approximate four-to-six-week mission, it will travel 280,000 miles from Earth and 40,000 beyond the far side of the Moon, demonstrating the performance of both Orion and the SLS rocket on their maiden flight, gathering important engineering data. Orion will stay in space longer than any spacecraft for astronauts has without docking to a space station and return home faster and hotter than ever before.

SLS and Orion will blast off from Launch Complex 39B at NASA's modernized spaceport at Kennedy. The SLS rocket is designed for missions beyond low-Earth orbit carrying crew or cargo to the Moon and beyond. The first SLS vehicle, called Block 1, will create more than 8.8 million pounds of thrust during launch and ascent and will be capable of delivering more than 27 metric tons (59,525 pounds) to orbits beyond the Moon. Propelled by a pair of five segment boosters and four RS-25 engines, the rocket will reach the period of greatest atmospheric force within ninety seconds.

After jettisoning the boosters, service module panels, and launch abort system, the core stage engines will shut down and the core stage will separate from the spacecraft, leaving Orion attached to the interim cryogenic propulsion stage (ICPS) that will propel it toward the Moon.

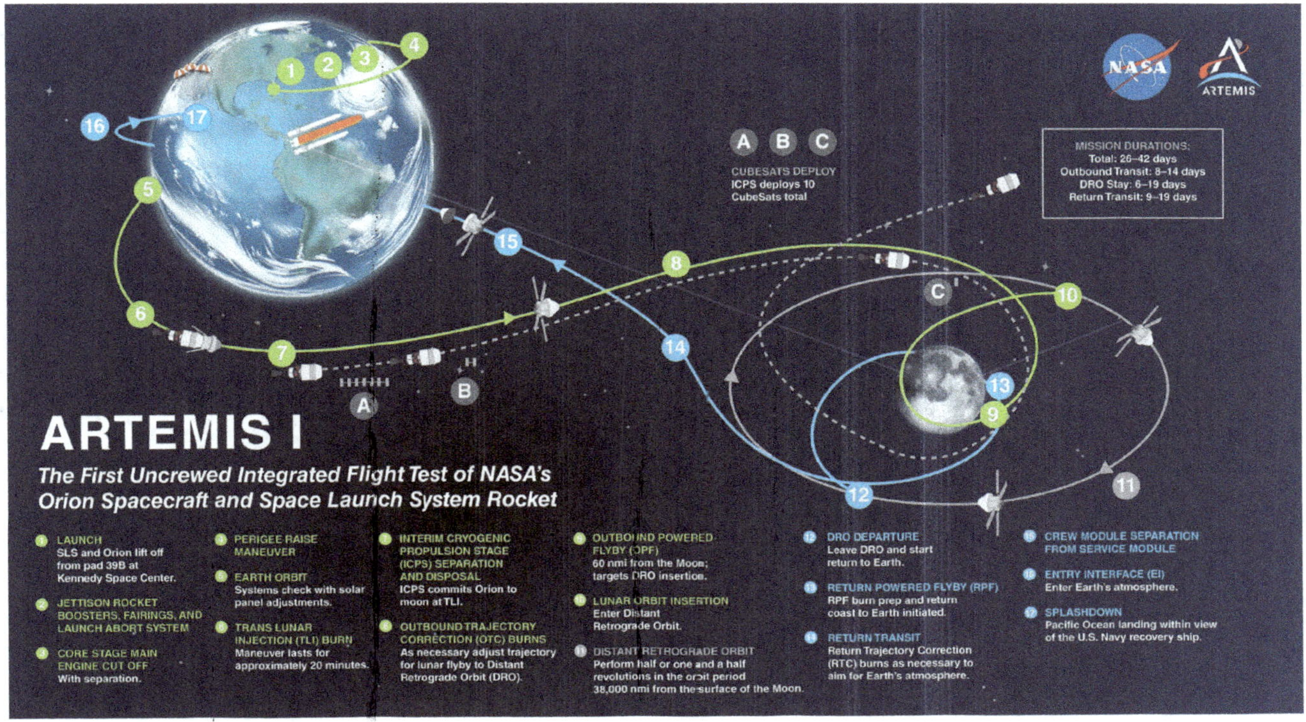

As the spacecraft makes an orbit of Earth, it will deploy its solar arrays and the ICPS will give Orion the big push it needs to leave Earth's orbit and travel toward the Moon. From there, Orion will separate from the ICPS approximately two hours after launch. The ICPS will then deploy several small satellites, known as CubeSats, to perform several experiments and technology demonstrations that will improve our knowledge of the deep space environment.

As Orion continues on its path from Earth orbit to the Moon, it will be propelled by a service module provided by ESA (European Space Agency). The service module supplies the spacecraft's main propulsion system and power, and houses air and water for astronauts on future missions. Orion will pass through the Van Allen radiation belts, and fly past the Global Positioning System (GPS) satellite constellation and above communication satellites in Earth orbit. To communicate with mission control at NASA's Johnson Space Center in Houston, Orion will switch from NASA's Tracking and Data Relay Satellites system and connect through the Deep Space Network. From here, Orion will continue to demonstrate its unique design in navigation, communication, and operations in a deep space environment.

The outbound trip to the Moon will take several days, during which time engineers will evaluate the spacecraft's systems and, as needed, correct its trajectory. Orion will fly about 60 miles above the surface of the Moon at its closest approach, and then use the Moon's gravitational force to propel Orion into a distant retrograde, or opposite, orbit about 40,000 miles past the Moon. This is 30,000 miles farther than Apollo travelled during Apollo 13 and the farthest in space any spacecraft built for humans has ever flown.

The spacecraft will stay in that orbit for at least six days to collect data and allow mission controllers to assess the performance of the spacecraft. During this period, Orion will travel in a direction around the Moon opposite from the direction the Moon travels around Earth.

For its return trip to Earth, Orion will do another close flyby of the Moon that takes the spacecraft within about 60 miles of the lunar surface, then use another precisely timed engine firing of the service module in conjunction with the Moon's gravity to accelerate back toward Earth. This maneuver will set the spacecraft on its trajectory to enter our planet's atmosphere traveling around 25,000

Orion and Artemis I Fun Facts

» Orion will be the spacecraft for crewed missions during Artemis missions. The name for the Artemis program comes from the twin sister of Apollo and   goddess of the Moon in Greek mythology. Artemis now personifies NASA's path to the Moon
and beyond.

» Orion returns to Earth at a speed of about 25,000 mph from its missions. At this speed, Orion could travel from Los Angeles to New York City in 6 minutes, while a normal flight on commercial airlines takes 5.5 hours.

» When it comes back to Earth, Orion will travel:

  » 17 times the speed of a fighter jet

  » 30 times the speed of sound

  » 350 times as fast as a cheetah runs

» Orion will go about 40,000 miles past the Moon on its Artemis I mission. That's:

  » 15 times the distance from Maine to San Diego past the Moon

  » 17 times the length of the Mississippi river past the Moon

» When it lifts off for Artemis I, Orion will weigh 72,000 pounds, as much as:

  » 4.5 elephants

  » 8.5 hippos

  » 13 giraffes

  » 50 cows

  » 80 horses

  » 200 pandas

mph and producing temperatures of approximately 5,000 degrees Fahrenheit, testing the heat shield's performance.

The mission will end with a test of Orion's capability to return safely to Earth as the spacecraft makes a precision landing within eyesight of the recovery ship off the coast of San Diego. Following splashdown, Orion will remain powered for a period of time as divers from the U.S. Navy and operations teams from NASA's Exploration Ground Systems approach in small boats from the waiting Naval recovery ship. The divers will briefly inspect the spacecraft for hazards and hook up tending and tow lines, then engineers will tow the capsule into the well deck of the recovery ship to bring the spacecraft home.

Launch Date	2022
Launch Site	Kennedy Space Center, Florida, Launch Pad 39B
Launch Vehicle	Space Launch System Block 1
Orion Gross Liftoff Weight	72,000 lbs.
Trans-Lunar Injection Mass	53,000 lbs.
Post-Trans Lunar Injection Mass	51,500 lbs.

Secondary Payloads

The payload mass capability of the SLS and the unused volume of the Orion stage adapter that connects the interim cryogenic propulsion stage (ICPS) to the spacecraft provide a rare opportunity for small, low-cost science and technology experiments to be deployed into deep space. These secondary payloads, known as CubeSats, are not much larger than a shoebox but contain science and technology investigations or technology demonstrations that help pave the way for future, deep space human exploration. International space agency partners and universities are involved with several of the CubeSat payloads.

The Artemis I CubeSats are 6U in size. One U – or unit – is 10 cm x 10 cm x 10 cm. The Artemis I payloads are limited to about 25 pounds (11.3 kg) each. Several of the CubeSats chosen to fly on Artemis I are lunar-focused and

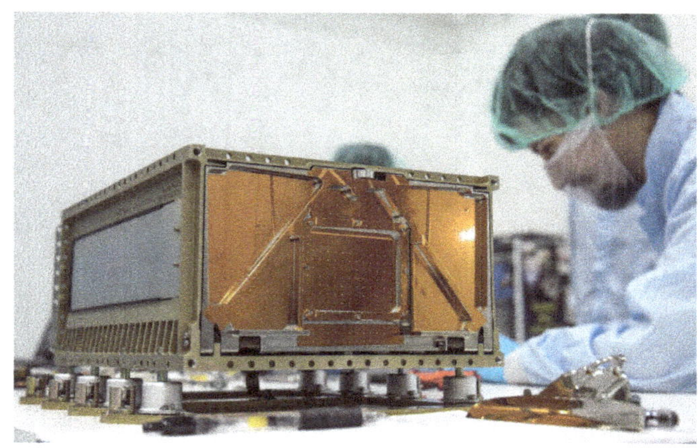

▲ Teams work on the final processing of the payloads that will fly aboard Artemis I. Housed within the Orion stage adapter, the satellites – called CubeSats – are roughly the size of a large shoe box and weigh no more than 30 pounds.

may help NASA address Strategic Knowledge Gaps (SKGs) to inform research strategies and prioritize technology development of human and robotic exploration. Other missions will be testing innovative propulsion technologies, studying space weather, analyzing the effects of radiation on organisms, and providing high resolution imagery of the Earth and the Moon. Some of the CubeSats are competing in the Cube Quest Challenge, vying for prizes for accomplishing such goals as farthest communication to Earth from space. The CubeSats will be deployed after Orion separates from the Orion stage adapter and ICPS and is a safe distance away. Each payload will be ejected with a spring mechanism from dispensers installed on the Orion stage adapter.

Artemis I Secondary Payload Fun Facts

» Deployment opportunities are along the upper stage disposal trajectory.

» The Secondary Payload Deployment System includes an avionics unit, mounting brackets, cable harnesses, and a vibration mitigation system.

» Weight Limit: 25 lbs. (11.3 kilograms) each

» Size: 6U (4.4' x 9.4" x 14.4") Equivalent to six 10 cm square units

Artemis I manifested payloads are listed below with their provider, area of exploration:

Moon

Lunar IceCube - Morehead State University, Kentucky Searching for water in all forms and other volatiles with an infrared spectrometer

LunaH-Map – Arizona State University, Arizona
Creating higher-fidelity maps of near-surface hydrogen in craters and other permanently shadowed regions of the lunar South Pole with neutron spectrometers

OMOTENASHI – JAXA, Japan
Developing the world's smallest lunar lander and studying the lunar environment

LunIR – Lockheed Martin, Colorado
Performing advanced infrared imaging of the lunar surface

Earth

EQUULEUS – University of Tokyo/JAXA, Japan
Imaging the Earth's plasmasphere for a better understanding of Earth's radiation environment from Earth-Moon LaGrange 2 point

Other Missions

BioSentinel – Ames Research Center, California
Using single-celled yeast to detect, measure and compare the impact of deep- space radiation on living organisms over a long period of time

ArgoMoon – European Space Agency/ASI, ArgoTec, Italy
Observing the interim cryogenic propulsion stage with advanced optics and software imaging system

Sun

CuSP – Southwest Research Institute, Texas
Measuring particles and magnetic fields as a space weather station

Centennial Challenges

Team Miles – Florida
Demonstrating propulsion using plasma thrusters and competing in NASA's Deep Space Derby

Asteroid

NEA Scout – Marshall Space Flight Center, Alabama
Traveling by solar sail to a near-Earth asteroid and taking pictures and other characterizations of its surface

Space Biology Experiments

Four space biology investigations will be carried in a container stored in the crew compartment of the Orion capsule for the duration of the mission, as it passes through the Van Allen Belts on the way to the Moon and again on the way back to Earth. The experiments will study DNA damage and protection from radiation for missions to the Moon, where radiation exposure will be roughly twice what it is on the International Space Station.

Artemis I provides an opportunity to study effects from the combined environment of space radiation and microgravity. Beyond low-Earth orbit, cosmic radiation is trapped in the Van Allen Belts that are part of Earth's magnetosphere and can be dangerous to humans and other living organisms, as well as sensitive electronics. The Moon is about 240,000 miles from the Earth, which is about 1,000 times more than the distance from Earth to the International Space Station and offers a true deep space radiation environment. All specimens will be returned to the researchers for post-flight analyses after the spacecraft returns. NASA selected investigators from four institutions for awards totaling ~$1.6 million in fiscal years 2019-2022.

Federica Brandizzi, Ph.D., Michigan State University, Life beyond Earth

Effect of space flight on seeds with improved nutritional value - This study will characterize how spaceflight effects nutrient stores in plant seeds with the goal of gaining new knowledge that will help increase the nutritional value of plants grown in spaceflight.

Timothy Hammond, Ph.D., Institute For Medical Research, Inc., Fuel to Mars

This research will include studies with the photosynthetic algae, Chlamydomonas reinhardtii, to identify important genes that contribute to its survival in deep space.

Zheng Wang, Ph.D., Naval Research Laboratory, Investigating the Roles of Melanin and DNA Repair on Adaptation and Survivability of Fungi in Deep Space

Researchers will be using the fungus Aspergillus nidulans to investigate radioprotective effects of melanin and the DNA damage response.

Luis Zea, Ph.D., University of Colorado, Boulder, Multi-Generational Genome-Wide Yeast Fitness Profiling Beyond and Below Earth's van Allen Belts

This investigation will use yeast as a model organism to identify genes that help organisms adapt to the conditions of both deep spaceflight on the Artemis I mission, and of low-Earth orbit on the space station.

NASA awarded these grants under the NASA Research Announcement NNH18ZTT001N-EM1 "Appendix A: Orion Exploration Mission-1 Research Pathfinder for Beyond Low-Earth Orbit Space Biology Investigations" which is part of the Research Opportunities in Space Biology (ROSBio) Omnibus. The Space Biology Program is managed by the Space Life and Physical Sciences Research and Applications Division in NASA's Human Exploration and Operations Mission Directorate at the agency's headquarters in Washington, D.C.

Matroshka AstroRad Radiation Experiment (MARE)

Space radiation is one of the most significant hazards crews face on missions to the Moon and beyond. In 2018, NASA signed an agreement with Lockheed Martin Advanced Programs, the Israel Space Agency (ISA), and the German Aerospace Center (DLR) for an experiment to test the AstroRad radiation protection vest on Artemis I. The investigation, called the Matroshka AstroRad Radiation Experiment (MARE), will provide valuable data on radiation levels during missions to the Moon while testing the effectiveness of the new vest.

The Artemis I mission will not carry crew, but two identical manikin torsos equipped with radiation detectors. The manikins, called phantoms, are manufactured from materials that mimic human bones, soft tissues, and organs of an adult female. One phantom, named Zohar, will wear the AstroRad vest, while the other, named Helga, will not. Female forms were chosen because women typically have greater sensitivity to the effects of space radiation, but the AstroRad vest is designed to protect both males and females.

The phantoms have a three-centimeter grid embedded throughout the torsos that will enable scientists to map internal radiation doses to areas of the body that contain critical organs.

With two identical torsos, scientists will be able to determine how well the new vest might protect crew from solar radiation, while also collecting data on how much radiation astronauts might experience inside Orion on a lunar mission – conditions that cannot be recreated on Earth.

Orion was designed to protect both humans and hardware during radiation events on Artemis missions. For example, in the event of a solar flare, Orion's crew can take cover in the central part of the crew module, between the floor and the heat shield, inside two large stowage lockers. Using the stowage bags that were in the lockers, they will create a shelter around themselves, putting as much mass between them and the radiation source as possible and making it harder for solar particles to travel through. With a protective vest to help block solar energetic particles, crews could continue working during critical mission activities in spite of a solar storm.

ISA will provide the AstroRad vest for the mission, which was developed by StemRad, an Israeli company, in collaboration with Lockheed Martin. DLR will provide the phantoms and most of the radiation detectors, with further contributions by universities from around the world.

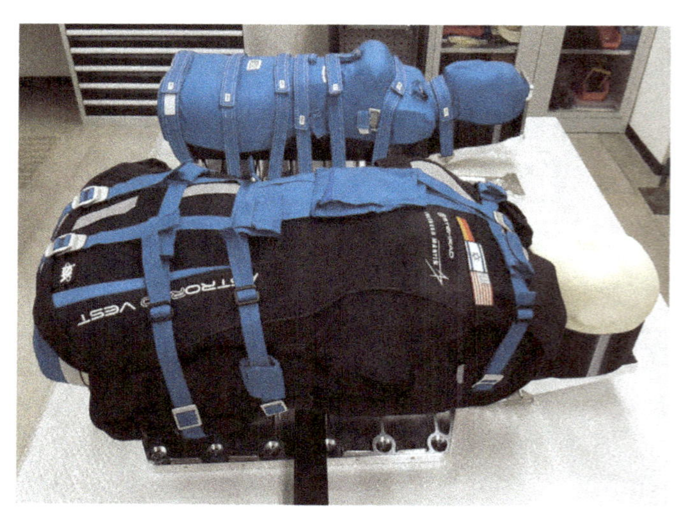

During NASA's Artemis I mission, two identical 'phantom' torsos named Helga and Zohar will be equipped with radiation detectors while flying aboard Orion to measure the effects of radiation in space, and to test for protection. Zohar will wear the vest, while Helga will not.

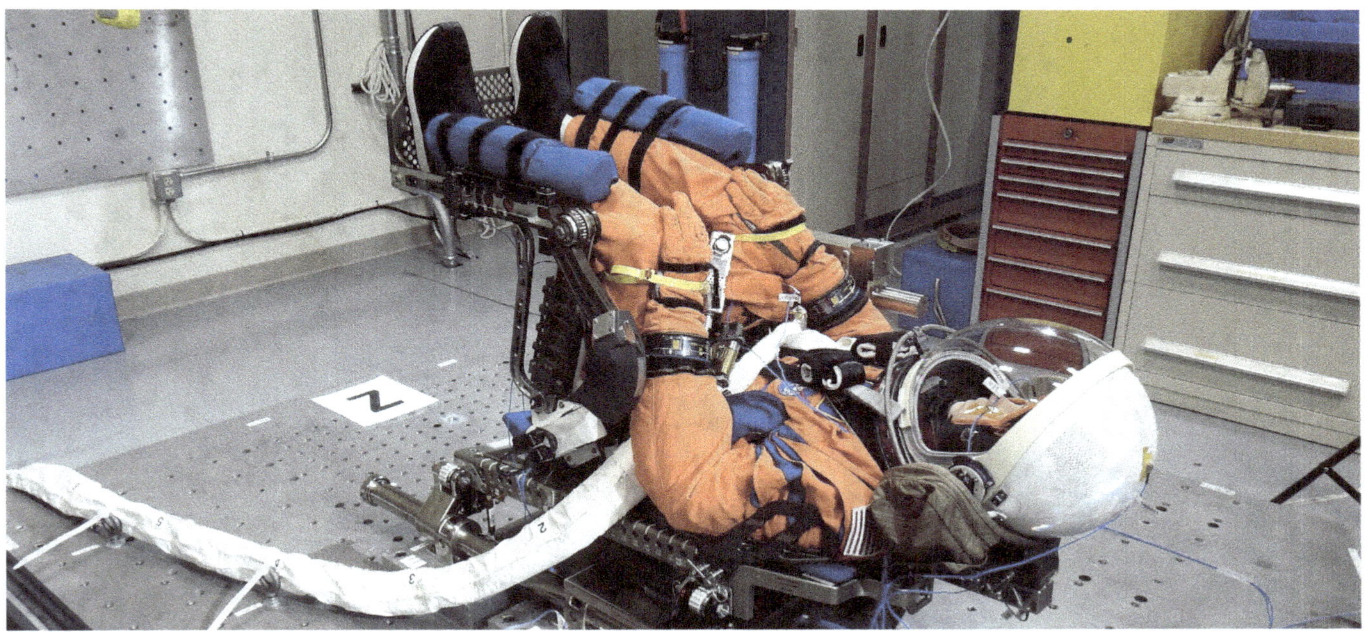

▲ The manikin flying on Artemis I will occupy the commander's seat inside Orion, be equipped with two radiation sensors, and wear a first-generation Orion Crew Survival System suit – a spacesuit astronauts will wear during launch, entry, and other dynamic phases of their missions.

Commander Moonikin Campos

The "Moonikin," a male-bodied manikin previously used in Orion vibration tests, will fly inside Orion on Artemis I to collect data for engineers on the ground to evaluate and fine tune crew conditions during missions to the Moon.

The Moonikin received the official name "Commander Moonikin Campos" as the result of a competitive bracket contest honoring NASA figures, programs, or astronomical objects. The name Campos is a dedication to Arturo Campos, a key player in bringing Apollo 13 safely back to Earth. The contest received more than 300,000 votes.

Campos will occupy the commander's seat inside Orion and wear an Orion Crew Survival System suit— the same spacesuit that Artemis astronauts will use during launch, entry, and other dynamic phases of their missions.

Campos will be equipped with two radiation sensors and have additional sensors under its headrest and behind its seat to record acceleration and vibration data throughout the mission. Data from the Moonikin's experience will be used to help NASA protect astronauts during future crewed Artemis missions.

Five additional accelerometers inside Orion will provide data for comparing vibration and acceleration between the upper and lower seats. As Orion splashes down in the Pacific Ocean, all accelerometers will measure impact on these seat locations for comparison to data from water impact tests at NASA's Langley Research Center in Virginia to verify accuracy of pre-flight models.

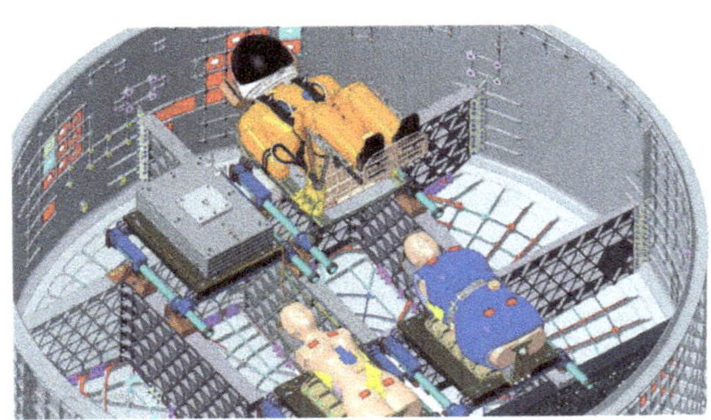

On Artemis I, Commander Moonikin Campos will occupy the commander's seat inside Orion, while phantom torsos Zohar and Helga will occupy the passenger seats.

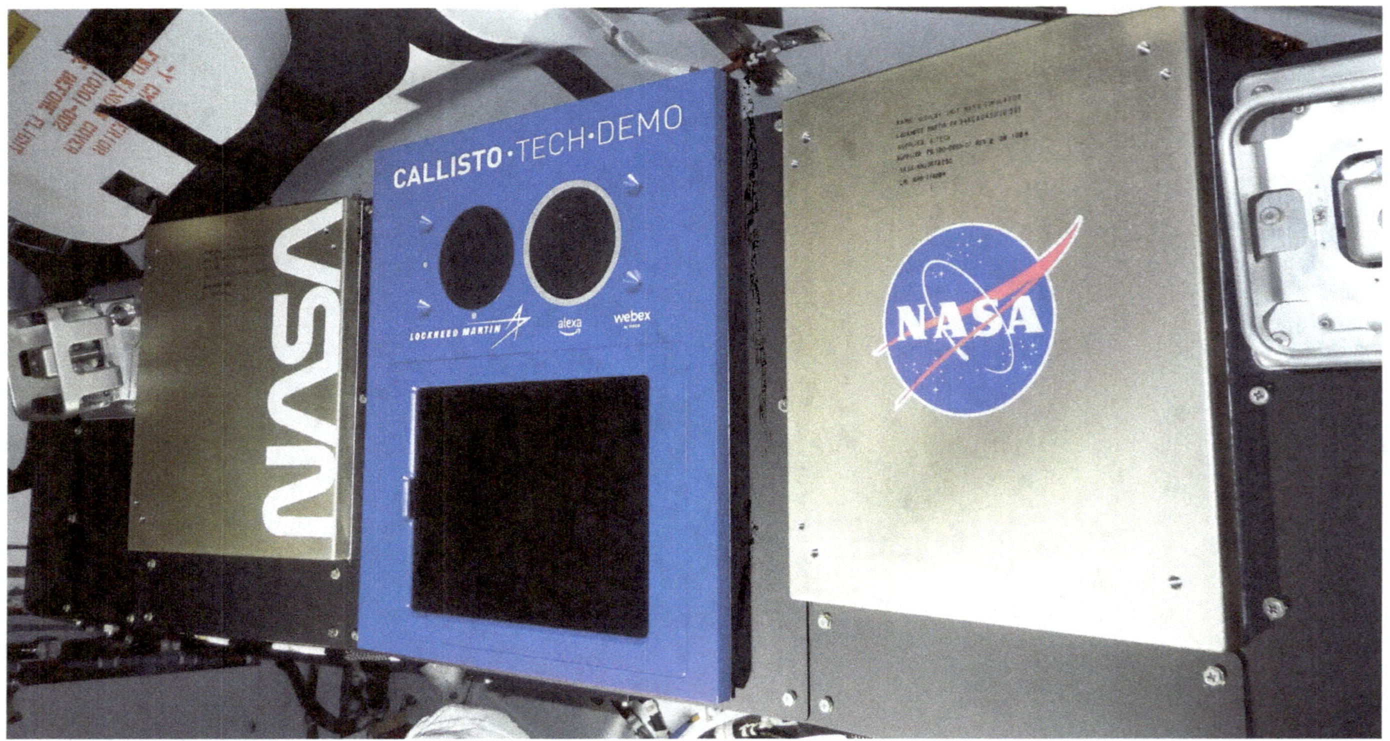

▲   Flying on NASA's Orion spacecraft during the uncrewed Artemis I mission will be Callisto, a technology demonstration developed through a reimbursable space act agreement with Lockheed Martin. Lockheed Martin has partnered with Amazon, and Cisco to bring the Alexa digital assistant and Webex video collaboration aboard Orion's first flight test in deep space.

Callisto

Callisto is a technology demonstration developed through a reimbursable space act agreement with Lockheed Martin. Lockheed Martin has partnered with Amazon and Cisco to bring the Alexa digital assistant and Webex video collaboration aboard Orion's first flight test in deep space.

Named after a mythological Greek goddess and one of Artemis' hunting attendants, Callisto is meant to show how commercial technology could assist future astronauts on deep space missions.

The payload will demonstrate how astronauts and flight controllers can use human-machine interface technology to make their jobs simpler, safer and more efficient, and advance human exploration in deep space.

The industry-funded payload will be located on Orion's center console and includes a tablet that will test Webex by Cisco video conferencing software to transmit video and audio from the Mission Control Center at Johnson, and custom-built hardware and software by Lockheed Martin and Amazon that will test Alexa, Amazon's voice-based virtual assistant, to respond to the transmitted audio.

Artemis II

Artemis II is the first Space Launch System (SLS) and Orion flight with astronauts. With confidence based on the Artemis I mission and the thousands of hours put into the prior flight and testing, this mission will send astronauts on an approximately 10-day mission where they will set a record for the farthest human travel beyond the far side of the Moon.

The mission will launch a crew of four astronauts from Kennedy on a Block 1 configuration of the SLS rocket. The flight profile is called a hybrid free return trajectory. Orion will perform multiple maneuvers to raise its orbit around Earth and eventually place the crew on a lunar free return trajectory in which Earth's gravity will naturally pull Orion back home after flying around the Moon.

The initial launch will be similar to Artemis I as SLS lofts Orion into space and then jettisons the boosters, service module panels, and launch abort system, before the core stage engines shut down and the core stage separates from the upper stage and the spacecraft. With crew aboard this

mission, Orion and the upper stage, called the interim cryogenic propulsion stage (ICPS), will then orbit Earth twice to ensure Orion's systems are working as expected while still close to home.

The spacecraft will first reach an initial orbit flying in the shape of an ellipse, at a low altitude (perigee) of about 115 miles by a high altitude (apogee) of 1,800 miles. The orbit will last a little over 90 minutes and include an initial firing of the ICPS to maintain Orion's path. After the first orbit, the ICPS will raise Orion to a high-Earth orbit. This maneuver will enable the spacecraft to build up enough speed for the eventual push toward the Moon. The second, larger orbit will take approximately 42 hours, with Orion flying in an ellipse between about 235 and 68,000 miles above Earth.

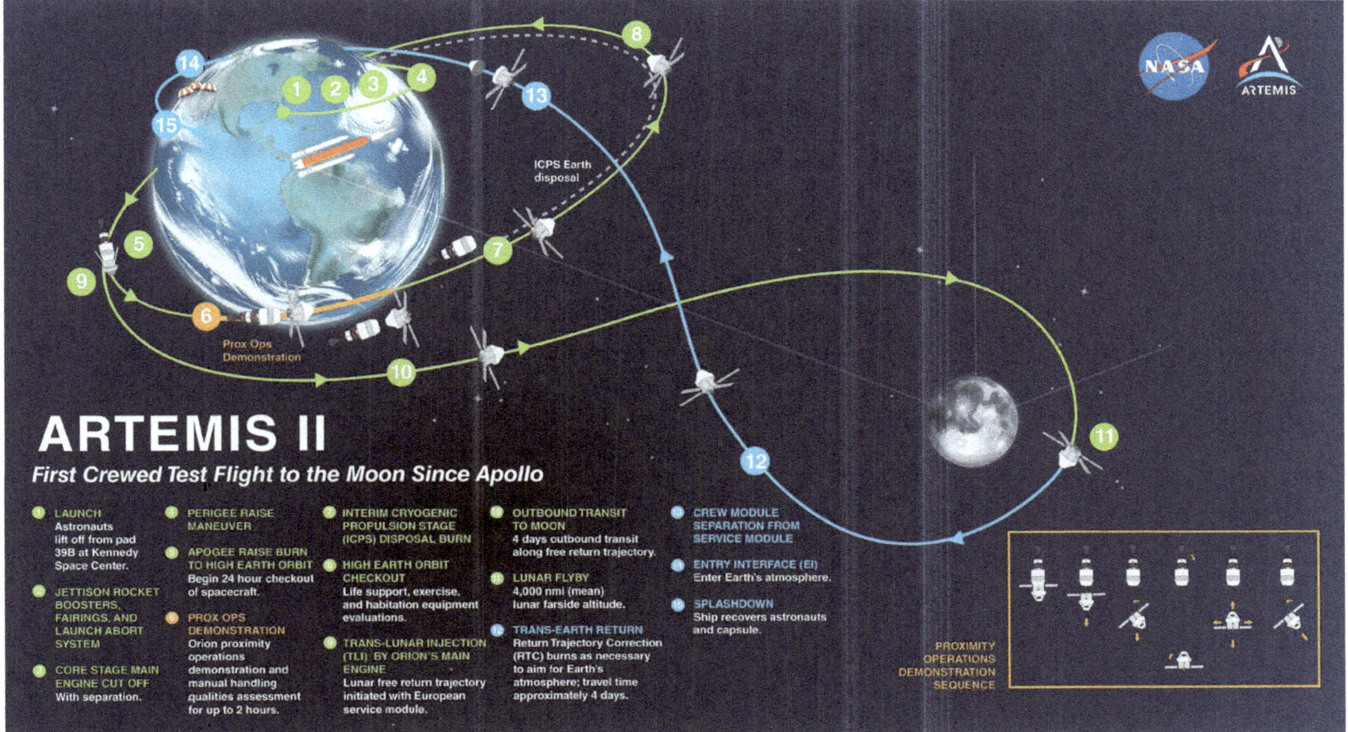

After the burn to enter high-Earth orbit, Orion will separate from the ICPS. The expended stage will have one final use before disposal through Earth's atmosphere—the crew will use it as a target for a proximity operations demonstration. During the demonstration, mission controllers in Houston will monitor Orion as the astronauts transition the spacecraft to manual mode and pilot Orion's flight path and orientation. The crew will use Orion's onboard cameras and the view from the spacecraft's windows to line up with the ICPS as they approach and back away from the stage to assess Orion's handling qualities and

related hardware and software. This demonstration will provide performance data and operational experience that cannot be readily gained on the ground in preparation for critical rendezvous, proximity operations, and docking, as well as undocking operations in lunar orbit beginning on Artemis III.

Following the proximity operations demonstration, the crew will turn control of Orion back to mission controllers at Johnson and spend the remainder of the orbit verifying spacecraft system performance in the space environment.

Technicians at NASA's Neil Armstrong Operations and Checkout Building at Kennedy Space Center in Florida move the Artemis II crew module from its clean room environment to an external work stand, where the environmental control and life support system and propulsion system components are assembled before installing them on the crew module.

In Bremen, Germany, the Airbus Space team prepares the Orion European Service Module-2 for shipment to Kennedy Space Center on Oct. 7, 2021.

While still close to Earth, the crew will assess the performance of the life support system's capabilities and ensure readiness for the lunar flyby portion of the mission. Orion will also briefly fly beyond the range of GPS satellites and the Tracking and Data Relay Satellites of NASA's Near Space Network to allow an early checkout of the agency's Deep Space Network communication and navigation capabilities.

After completing checkout procedures, Orion will perform the next propulsion move called the trans-lunar injection (TLI) burn. With the ICPS having done most of the work to put Orion into a high-Earth orbit, the service module's main engine will fire to perform the TLI burn that will propel Orion on a path toward the Moon. The TLI burn will send crew on an outbound trip of about four days
to lunar orbit, with the service module performing orbit travel 40,000 miles around the far side of the Moon, where they will ultimately create a figure eight extending about 280,000 miles from Earth before Orion returns home.

On the remainder of the trip, astronauts will continue to evaluate the spacecraft's systems, including demonstrating return operations, practicing emergency procedures, and testing the radiation shelter, among other activities.

With a return trip of about four days, the mission is expected to last just over 10 days. Instead of requiring propulsion on the return, the fuel-efficient free return trajectory harnesses the Earth-Moon gravity field, ensuring that—after its trip around the far side of the Moon—Orion will be pulled back naturally by Earth's gravity for the remaining portion of the mission.

During Orion's return to Earth, the crew module separates from the service module and begins entry, traveling nearly 25,000 mph. A series of drogue and pilot parachutes will slow Orion down from Mach 32 to zero in approximately 20 minutes, ending a mission that will exceed 620,000 miles. Recovery forces, already positioned at the target landing zone, will be ready to recover the crew from the Pacific Ocean.

Launch Site	Kennedy Space Center, Florida, Launch Pad 39B
Launch Vehicle	Space Launch System Block 1
Orion Gross Liftoff Weight	78,000 lbs.
Trans-Lunar Injection Mass	58,500 lbs.
Post-Trans Lunar Injection Mass	57,500 lbs.

Artemis III

Artemis III will fly the first woman and first person of color to the Moon. This flight will be the culmination of rigorous testing and more than two million miles accumulated in space on NASA's deep space transportation systems during Artemis I and II.

On this mission, Orion and its crew will travel to the Moon, this time docking directly with a Human Landing System (HLS) in lunar orbit, which will take the astronauts to the Moon's surface.

The agency's powerful Space Launch System rocket will launch four astronauts aboard the Orion spacecraft for their multi-day journey to lunar orbit. There, two crew members will transfer to the HLS for the final leg of their journey to the surface of the Moon.

NASA intends to implement a competitive procurement for sustainable crewed lunar surface transportation services that will provide human access to the lunar surface using the Gateway on a regularly recurring basis beyond the initial crewed demonstration mission.

The exact landing site for Artemis III astronauts depends on several factors, including the specific science objectives and the launch date. High-resolution data received from NASA's Lunar Reconnaissance Orbiter has provided incredible views and detailed mapping of the lunar surface, including changes in lighting throughout the year. The agency is working with the global science community to study different regions that provide key desired traits: access to significant sunlight, which provides minimal temperature variations and potentially the only power source; continuous line-of-sight to Earth for mission support communications; mild grading and surface debris for safe landing and walking or roving mobility; and close proximity to permanently shadowed regions, some of which are believed to contain resources such as water ice.

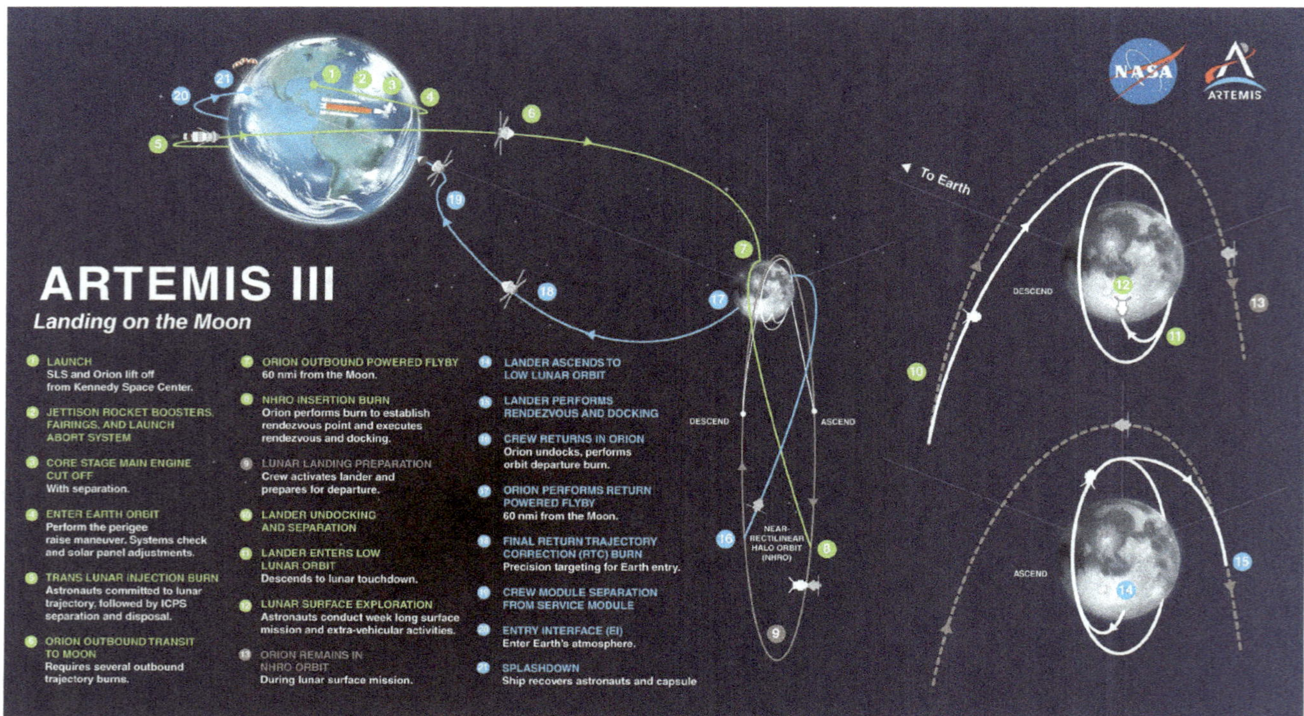

At NASA's Michoud Assembly Facility in New Orleans, technicians from Orion prime contractor Lockheed Martin weld together three cone-shaped panels on Orion's crew module for the Artemis III mission that will land the first woman and next man on the Moon.

In addition to two crew, the HLS will carry up to 220 pounds of science tools and equipment to the surface, with the goal of returning up to 87.5 pounds of samples. Also, our Commercial Lunar Payload Services (CLPS) providers may be used to deliver pre-emplaced science instruments and equipment for use by our first human return crew while exploring on the lunar surface.

After approximately a week exploring and completing this historic expedition on the lunar surface, the crew will board the lander for their short trip back to orbit where they will return to Orion and their colleagues. With their pristinely preserved samples from the Moon, the crew will prepare for the trip back to Earth.

Launch Site	Kennedy Space Center, Florida, Launch Pad 39B
Launch Vehicle	Space Launch System Block 1
Orion Gross Liftoff Weight	78,000 lbs.
Trans-Lunar Injection Mass	58,500 lbs.

Johnson Space Center

NASA's Orion program is managed at Johnson Space Center (JSC), where a team of engineers oversees the design, development and testing of the spacecraft, as well as spacecraft manufacturing taking place across the country and in Europe.

JSC has led human space exploration for more than half a century. As the nucleus of the nation's astronaut corps, the home to the Orion crew, and the iconic Mission Control Center, JSC plays a pivotal role in human spaceflight and enhancing technological and scientific knowledge to benefit humankind. From the Mercury, Gemini, Apollo and Space Shuttle programs to the International Space Station, Commercial Crew and Orion programs, the center has been at the forefront of America's human spaceflight.

Mission Control Center

Since 1965, mission control has been the helm of America's human spaceflight. NASA's Christopher C. Kraft Mission Control Center (MCC) at the Johnson Space Center in Houston is the primary facility where flight controllers command and control human spacecraft missions. Since International Space Station assembly began in 1998, the center has become a focal point for human spaceflight worldwide. The teams that work in mission control have been vital to every U.S. human spaceflight since the Gemini IV mission in 1965, including the Apollo missions that took humans to the Moon, the more than 110 space shuttle flights since 1981 and today's commercial crew flights.

MCC and Artemis I

MCC is comprised of several flight control rooms (FCR), including FCR-1, FCR-2 and the Red, White, and Blue FCRs. Work to transform the White FCR from its shuttle legacy configuration into a modern mission control configuration to support 21st century missions, known as the MCC-21 effort, began in January 2013. The upgraded White FCR is ultimately intended to serve as the mission control for flights of NASA's Orion spacecraft on missions to deep space destinations.

MCC is the facility from which flight operations personnel will remotely monitor and operate the Orion spacecraft and receive data from Orion and the Space Launch System (SLS) during Artemis missions. During Artemis I, as Orion travels toward the Moon over the course of several days, flight controllers will command Orion from the ground six times to correct its trajectory to ensure the spacecraft can fly by the Moon at the correct time and place. This includes trajectory corrections that enter Orion into the planned lunar orbit, exit that orbit, and then return to Earth to splashdown in the Pacific Ocean.

The flight control team has been training for the Artemis I mission since 2019, by constantly refining and

▶ In mission control at Johnson Space Center in Houston, flight controllers simulate part of Orion's uncrewed flight to the Moon for Artemis I.

command, and control Orion, and continues to do so until several weeks before the mission. This includes performing numerous simulations at MCC of the various parts of Orion's journey. Flight controllers at MCC train and prepare by simulating Orion's launch through outbound powered flyby to the Moon, including the trans-lunar injection (TLI) burn that sends the spacecraft out of Earth orbit and toward the Moon. MCC also simulates Orion's orbit around the Moon and return powered flyby from the Moon through entry, descent, landing, and recovery, including the final trajectory corrections and burns Orion will need to enter the atmosphere and splashdown in the Pacific Ocean.

Simulations and testing have included joint operations between Lockheed Martin and the NASA Flight Operations Directorate, with NASA teams doing real-time monitoring and commanding of the Integrated Test Lab (ITL) Orion vehicle in Denver from the MCC in Houston. The ITL Orion vehicle replicates Orion's computer, wiring, and avionics systems configurations. Testing has also been done with a low-fidelity Orion at the Exploration Development Lab in Houston as well as with the actual Orion spacecraft.

Engineers have also tested the Orion communications system to ensure the spacecraft and MCC can flawlessly communicate and send data through NASA's satellite networks in space and on the ground. To communicate during Artemis I, MCC and Orion will use NASA's Near Space Network, Tracking and Data Relay Satellite (TDRS) services, and Deep Space Network. The Near Space Network consists of Direct-to-Earth services, which use a worldwide network of ground stations to provide communications and navigation services during launch, as well as navigation services at various points on Artemis I. Tracking and Data Relay Satellite (TDRS) services provide near-continuous communications services to spacecraft near Earth, and play a critical role during launch and low-Earth orbit phases of Artemis I as well as during entry into Earth's atmosphere, splashdown, and recovery. The Deep Space Network will handle Artemis I communications beyond Near Space Network coverage, en route to and in orbit around the Moon.

Teams perform communication testing between NASA's Mission Control Center in Houston and a lab at Orion prime contractor Lockheed Martin's facility near Denver that replicates Orion's computer, wiring and avionics systems configurations.

MCC has verified these communication systems work with Orion during tests of different Artemis I scenarios, including pre-launch countdown through the point at which Orion data is relayed through the DSN, operations in lunar orbit, handover between the DSN and the Near Space Network during Orion's trajectory from the Moon back toward Earth, and post-splashdown operations. Previous testing also verified communications through the Near Space Network satellites and ground station at Cape Canaveral, and also included support from personnel at the Huntsville Operations Support Center, also known as the HOSC, at NASA's Marshall Space Flight Center, to verify they can receive data from the SLS rocket. Engineers at the SLS Engineering Support Center at Marshall also conducted voice tests to ensure teams across the country, including flight controllers in Houston, launch controllers in Florida, and engineering teams at several locations including in Huntsville can communicate by voice.

Rapid Prototyping Lab

The focus of NASA's Rapid Prototyping Laboratory (RPL) located at the Johnson Space Center in Houston is to define and develop the whole suite of over 60 display formats and interactive controls for the Orion spacecraft.

The RPL team provides quick, creative, and efficient solutions to an array of issues that engineers and astronauts may face during missions. The team has used this approach to streamline the more than 2,000 switches used on the space shuttle into a new interface for the Orion cockpit, using just three display units, about 60 physical switches, two rotational hand controllers, two translational hand controllers, and two cursor control devices.

This new interface not only eliminates mass from Orion, but the software also makes it easier to operate the capsule by automatically generating dialogue boxes to help processes move along more efficiently. Astronauts on the space shuttle relied on nearly 200 pounds of operating manuals to help them fly the orbiter. On Orion, electronic procedures are programmed into the system to aid the crew in daily and emergency processes, making these large manuals of system operations obsolete and saving the crew time and space.

The RPL team prototypes these display formats as part of an integrated cockpit simulation hosted on various Orion cockpit mockups around the Johnson Space Center. The mockup is being used to provide a realistic environment for astronauts to evaluate the Orion cockpit and to practice Orion spacecraft operations. It is used to implement test scenarios to help decide on final design issues such as the types of controls during the different phases of flight, organization of the display formats and navigation, hardware selection and ergonomics, operational concepts, and display format specification and prototyping.

The Orion cockpit mockup includes three display units, hand controllers and cursor control devices for two operators, and seven switch panels set up in an "on your back" fashion to simulate ascent and entry into Earth's atmosphere. All the controls in the mockup are functional and connected to a simulation of the Orion spacecraft display suite. A half shell, recovered from another project,provides a representation of being inside a capsule. Additionally, an out-the-window video taken during EFT-1 is projected onto a screen outside the mockup shell, giving users an idea of what the view would be like on the way to and from space. Audio files from the flight test are synchronized with the video and played through a surround sound system,

Finally, vibration units, driven by the audio, are mounted to the chairs to highlight major events, such as booster separation and splashdown.

While simulations are being run on the prototype Orion crew interface display system, a mock mission control room is a few doors down, where flight controllers gain experience working and communicating with the astronauts using the crew-vehicle interface system in real time.

The team works closely with a diverse group of engineers, system experts, flight controllers, crew, human engineering, and Orion's prime contractor, Lockheed Martin, throughout the entire prototyping process to ensure the operating system complies with human spaceflight standards. A demo at the Orion cockpit mockup is also a popular item on VIP tours at the Johnson Space Center.

Systems Engineering Simulator/Reconfigurable Operational Cockpit

The Systems Engineering Simulator (SES) at NASA's Johnson Space Center is a real-time, crew-in-the-loop engineering simulator for the space station, Orion, and advanced programs. The SES provides the ability to test changes to existing space vehicles and flight software, test the interaction of a new vehicle system with existing systems, create models of new vehicles (that may or may not exist yet) for engineering analysis, and evaluate display and control concepts and modifications.

For Orion, the SES provides simulation of the vehicle with accurate six-degree-of-freedom equations of motion, the Orion guidance, navigation, and control system, and sensor and effectors; and allows for Orion engineering studies such as proof of concept, operational feasibility, and design assessment. The SES also provides mission support and evaluation through procedure development, training, and flight support.

▲ NASA Astronauts Stephanie Wilson, Jonny Kim, and Randy Bresnik take a look at the recently arrived Orion spacecraft simulator at the agency's Johnson Space Center in Houston on Dec. 8, 2020.

The Reconfigurable Operational Cockpit (ROC) is a reconfigurable cockpit mockup located within the SES Beta Dome, a 24-foot-diameter hemispherical dome upon which one continuous image is projected. This allows crewmembers and engineers to perform tests, evaluations, or training in a controlled cockpit environment while viewing a modeled external environment through the cockpit windows. Orion cockpit simulations have been performed with the ROC, which can simulate both seated and reclined Orion cockpit positions and can use the dome visual system to simulate out-the-window viewing.

Orion Mission Simulator

The Orion Mission Simulator (OMS) is a full-task mission trainer at NASA's Johnson Space Center used to train crew and flight control teams on Artemis missions. The high- fidelity vehicle simulations, which incorporate real flight software, along with extensive malfunction capabilities, provide these teams training on nominal flight activities as well as complex malfunction scenarios, including emergency events.

The OMS is comprised of several components: Orion and SLS Program-provided software simulations and supporting files, Flight Operations Directorate (FOD) training systems infrastructure, user support tools, network simulations, and crew station hardware.

The SOCRRATES-lite (Software Only CEV Risk Reduction Analysis and Test Engineering Simulator) and the ARTEMIS (Ares Real-Time Environment for Modeling, Integration, and Simulation) are used to model both Orion and SLS, respectively. Both are deployed in the FOD training systems infrastructure, which includes computational hardware and data network equipment. The infrastructure also provides a communications and tracking network simulation) of various space to ground networks and allows the vehicle simulations to communicate with the Mission Control Center (MCC) in a very flight-like manner.

The flight crew uses the crew station to interface with the vehicle simulations and with the MCC. The forward cabin of the crew station is accurate in scale to the flight vehicle for the pilot and commander, as well as all four windows with high fidelity simulated out-the-window scenes. A full-scale docking tunnel is also included and will be outfitted with a low fidelity hatch for rendezvous and proximity operation tasks. The crew station is outfitted with high-fidelity displays and controls and high-fidelity crew seats. A breathing air system is also included that will utilize high-fidelity Apollo fittings to supply air to crew, as well as a chilled water suit cooling system to support two suited crew members.

The OMS will offer the crew both visual and aural cues to reinforce data seen on the display units. Sound bites from recorded missions will be tied to parameters in the simulation: when a trigger is set, the appropriate sound bite will annunciate. Simulated window views will also be generated based on simulation parameters. These cues are important in providing the crew situational awareness.

For Artemis I, the SOCRRATES and ARTEMIS simulations are run independently, and the crew station is not used. For Artemis II, SOCRRATES and ARTEMIS will be integrated to run as a single stack simulation and will be integrated with the crew station. The simulations will be modified as required to support future missions.

Platform	288" Platform + 132" Stairs = 420" L x 318" D x 167" H
Crew Station	190.94" W x 184" L x 125.95" H, fixed-base, oriented horizontally with the seats looking forward.
Total Volume Inside Crew Station	1005 cubic feet
Overall Artemis II Crew Station Dimensions Including Platform	420" L x 318" D x 242.57" H

Space Vehicle Mockup Facility

The Space Vehicle Mockup Facility (SVMF) at Johnson Space Center provides spatially accurate flight vehicle mock-ups for engineering development, world class training for space flight crews and their support personnel, as well as high-fidelity hardware for real-time mission support. A major task of the SVMF is to support engineering and mission operations evaluations for the Orion Program. All mockups and part-task trainers are available to support troubleshooting on the ground any time problems develop on orbit in real time.

The SVMF houses a medium-fidelity mockup of the Orion capsule for support of future exploration. The items inside the mock-up of the Orion cabin are accurately placed within one inch of their location in the real flight vehicle. The mockup is currently used for development of components in the Orion cabin, assessments of interior layouts, and human-in-the-loop development and verification testing.

A variety of training is conducted in the SVMF mockup to prepare for emergency operations, on-orbit maintenance, photo/TV, stowage and handling, EVA/airlock operations, and scientific payload operations. The mockup is also used for a variety of engineering and operations evaluations such as on-board stowage evaluations, fit checks, and procedures development and verification.

Development and Testing > Johnson Space Center

At NASA's Johnson Space Center in Houston, testing takes place using the Orion mockup in the Space Vehicle Mockup Facility to ensure crew   can get out of Orion within two minutes to protect for a variety of failure scenarios that do not require the launch abort system to be activated.

▲   During testing using the Orion mockup in the Space Vehicle Mockup Facility at NASA's Johnson Space Center in Houston, engineers use fake smoke to imitate a scenario in which astronauts must exit the capsule when their vision is obscured.

Human-In-The Loop Evaluations

Human-in-the-loop evaluations, or HITL, are done at Johnson Space Center to make sure Orion's displays and controls are operable for all mission tasks and the living space is properly designed for astronauts of all sizes. HITL testing involves humans simulating and evaluating the tasks and interactions performed in the crew module, extending the design process beyond 2D modeling. The wide variety of testing in different Orion mockups reveals design and integration problems, and opportunities for cost efficient improvements.

Displays & Controls

Orion crew module display formats (i.e. software user interfaces) are frequently evaluated in HITL testing to ensure successful crew interaction for the breadth of tasks required. HITLs have informed the design and operation of the Orion displays and controls, including the hand controllers, switch interface panels, display format legibility (including during vibration), off-angle viewing, and design aspects such as switch guards, spacing, and strength to operate. This testing is performed in the most flight-like facilities possible in order to ascertain whether the crew will be able to safely pilot and control the vehicle given the expected avionics latency and system response, and for any anticipated contingency situations.

Seat Design

The design of seating in the crew module is tested to ensure that the seats can work for anyone from the 1st to 99th percentile of body anthropometries – from a 4-foot-10-inch, 94-pound female to a 6-foot-5-inch, 243-pound male. This includes testing the adjustable seat pans, foot plate, and arm rests of the seats and adjusting hand controller mounts to make sure that the smallest or largest of astronauts can reach all the controls.

Crew Vehicle Egress & Survival Operations

Testing takes place to make sure the crew can safely and efficiently egress the capsule and includes modifying hardware placement and operability. Timed emergency egress operations are also simulated to determine if time requirements are met using the emergency side hatch (or tandem hatch) opening for pre-launch or post-landing egress. Egress testing evaluates the crew's ability to exit the spacecraft from both the side hatch and docking hatch and includes many factors such as the ability to translate through the ocean to safely ingress a life raft in the event of an emergency on Orion.

Habitability & Environmental Systems

HITL testing is performed on crew systems such as the potable water dispenser and food warmer to ensure that the spacecraft is functional as a place to live as well as work. Testing includes the crew's ability to command changes to the atmosphere to maintain a safe and comfortable shirtsleeve environment, as well as the crew's ability to respond to emergencies such as a loss of cabin atmosphere or the response to a fire. There are many design controls to prevent these emergencies from occurring in the first place.

Net Habitable Volume

This HITL testing involves making sure there is adequate room for mission tasks and configurations inside the crew module. This includes testing configurations of the cabin for different mission phases, radiation shelter and sleeping bags, as well as testing exercise feasibility, suit donning and doffing, hygiene operations, and medical response. Net habitable volume is a key consideration in Orion cabin design given volumetric constraints with four crewmembers living and working in that volume throughout the mission.

Rendezvous, Proximity Operations, and Docking (RPOD) Operations

This testing evaluates tasks that will be performed in the Orion docking tunnel and vestibule area. This includes docking preparations, docked operations such as arrival and departure, and evaluating ease of operation of docking hardware. Subjects also simulate the steps that would be involved during the preparation and execution of rendezvous, proximity operations, and docking (RPOD) such as performing vestibule leak checks and removing and stowing the docking hatch in a protective cover. RPOD HITL testing also includes handling evaluations of ability to pilot Orion through RPOD fly-around and docking maneuvers. HITL testing also assesses contingency situations, such as the ability to conduct emergency hatch closure operations in the event of a micro-meteoroid/orbital debris strike, resulting in a cabin atmosphere depressurization.

Suited Operations

HITL evaluations take place to test astronauts' interactions with crew module interfaces while wearing the pressurized Orion Crew Survival System (OCSS) suit. Subjects don the OCSS suit and the suit is brought to pressure to perform various tasks and interactions with hardware. Suited operation HITLs are performed in either a flight- like environment, such as the Orion medium-fidelity mockup in the Space Vehicle Mockup Facility at NASA's Johnson Space Center in Houston, or in a pressurized glovebox facility for tasks where only pressurized glove interaction needs to be evaluated.

Speech Intelligibility

This series of HITLs evaluates the crew's ability to communicate with each other and with the Mission Control Center on the ground in various configurations that may be experienced during the mission. The HITL simulates the appropriate noise environment and sound attenuation and then rates the subject's ability to discern words and sounds. Speech intelligibility testing is critical to ensure that crew can communicate effectively during critical operations, including launch, entry, and in responding to emergencies.

Handling Qualities

This series of HITLs evaluates a pilot's ability to fly the spacecraft. Testing is done to evaluate how the vehicle responds to hand controller deflections during manual piloting, and to understand whether the total system (including displays and operator aids) allows the pilot to maintain control. Handling qualities HITLs are a critical aspect of human rating of the spacecraft, and they must be conducted in a very flight-like mockup with accurate displays, controls, and software in order to yield valid results.

Crew Module Uprighting System

When Orion splashes down in the ocean, it can settle in one of three positions: up, down or on its side. The crew module uprighting system (CMUS) deploys a series of five, bright orange airbags on the top of the capsule to flip Orion right side up in the event it lands and turns upside down.

The system will keep Orion upright and stable after splashdown in the ocean and for at least 24 hours in the waves, if necessary. It takes less than four minutes for the CMUS to upright the capsule. The capsule must be upright for crew module communication systems to operate correctly, and to help protect the health of the astronauts inside who are returning home from future deep space missions. Astronauts can suffer health impacts from extended time hanging upside down in their seat harnesses.

Engineers have implemented a series of improvements on the CMUS since Orion's Exploration Flight Test-1 in 2014. During the flight test, three of the system's five bags did not properly inflate. The spacecraft landed and remained upright in the water, however had the capsule landed upside down, the two functioning CMUS bags would likely not have been able to fully upright the capsule. Design improvements that have been made since the test include thickening the inner bladder of each round airbag to make it more durable, changing how the bags are packed, developing a hard enclosure for the packed bags, and improving manufacturing processes for better control and consistency.

The five bags that make up the CMUS are packed in hard containers and stored atop the capsule inside the structural gussets between the parachutes and other equipment. The areas around the CMUS airbags have been evaluated carefully for protrusions and sharp edges that could potentially puncture the bladder or snag the bags or tethers that keep the bags attached to the spacecraft. Changes to spacecraft components have been made to minimize the sharp edges and snag hazards, and full-scale testing has demonstrated the CMUS can work well alongside the other spacecraft equipment.

The five CMUS bags are inflated with helium gas that is stored in pressure vessels on the spacecraft throughout the mission. Each bag has an independent inflation system. After landing, the CMUS system is initiated and a valve is opened to allow helium to flow into the uprighting bags. As the gas fills the bags, they deploy from their containers and inflate to their full volume. The CMUS will deploy regardless of the landing position of the capsule.

Several full-scale uprighting tests have been performed with a mock-up of the Orion crew capsule. These tests demonstrated that the CMUS would still be able to perform as intended if any one of the five uprighting bags were to fail. This included seven tests in the calm water of the Neutral Buoyancy Lab (NBL), a giant pool at NASA's Johnson Space Center in Houston primarily used for astronaut training. The Johnson CMUS and NBL teams also successfully completed two tests off the coast of Galveston, Texas, in cooperation with the U.S. Coast Guard Cutter CYPRESS, Air Force personnel, and Texas A&M Galveston. An additional four tests were performed in the Atlantic Ocean, off the coast of Atlantic Beach, North Carolina, in cooperation with the
U.S. Coast Guard Station Fort Macon and the USCG Cutter MAPLE. These tests demonstrated performance in a natural wave environment and were instrumental in the certification of the CMUS. The nominal landing location for the Orion spacecraft is approximately 50 miles off the coast of San Diego.

▼ The Orion Crew Module Uprighting System (CMUS) and Neutral Buoyancy Laboratory team completed two successful sea tests off the coast of Galveston, Texas, from Dec. 1-3, 2018.

▲   Off the coast of Atlantic Beach, North Carolina, engineers tested the crew module uprighting system (CMUS) to ensure the capsule can be oriented right-side up once it returns from its deep space missions.

Neutral Buoyancy Lab CMUS Testing

Date	Test Configuration
6/23/2017	5 bags
2/6/2018	4 bags (180° out)
2/8/2018	4 bags (240° out)
2/14/2018	4 bags (300° out)
2/16/2018	4 bags (120° out)
6/25/2018	4 bags (240° out) – FRB retest
7/10/2018	4 bags (240° out) – FRB retest

At-Sea CMUS Testing

Date	Test Configuration	Location
12/1/2018	4 bags (120° out)	Galveston, TX
12/3/2018	4 bags (240° out)	Galveston, TX
3/7/2019 Q*	4 bags (240° out)	Atlantic Beach, NC
3/10/2019	4 bags (120° out)	Atlantic Beach, NC
3/12/2019 Q*	4 bags (60° out)	Atlantic Beach, NC
3/15/2019	4 bags (180° out)	Atlantic Beach, NC

White Sands Test Facility

NASA's White Sands Test Facility in New Mexico is a component of Johnson Space Center in Houston that tests and evaluates potentially hazardous materials, spaceflight components, and rocket propulsion systems for NASA centers, other government agencies, and commercial industry. Since the 1960s, White Sands has been an integral part of every United States human spaceflight mission, supporting programs associated with the safe exploration of space and leading the way in propulsion system testing. This includes testing of Orion's European Service Module, the powerhouse of the spacecraft that provides in-space propulsion, power and other astronaut life support systems, including consumables like water, oxygen, and nitrogen.

European Service Module Propulsion Testing

The propulsion qualification module (PQM) test article, a replica of the service module's propulsion subsystem, was delivered by ESA and Airbus to White Sands in 2017. The PQM was designed to verify the performance of the service module engines, propellant feed systems, and various other propulsion operations during expected and unexpected conditions. Testing of the PQM is crucial for ensuring that all engines and thrusters fire safely and accurately to get the spacecraft where it needs to go.

▼ At NASA's White Sands Test Facility near Las Cruces, New Mexico, a test using the propulsion qualification module (PQM) simulates what is referred to as an abort-to-orbit scenario, a scenario in which Orion's service module places the spacecraft in a safe orbit should a problem occur after the launch abort system has been jettisoned.

The PQM model is equipped with a total of 21 engines: one U.S. Space Shuttle Orbital Maneuvering System (OMS) engine, eight auxiliary thrusters, and 12 smaller thrusters produced by Airbus in Germany. The Orion spacecraft that will fly on Artemis missions has 33 engines in total, double the amount of RCS thrusters included in the PQM.

The full model is roughly a 15-foot cube made of stainless steel and provides the full components for testing the thrusters, fuel lines, and firing of Orion's engines. The propellant for the PQM is provided by four 2000-liter ground test tanks with 1-centimeter thick walls containing mixed oxides of nitrogen (MON) as oxidizer and monomethyl hydrazine (MMH) as fuel. Other elements of the PQM include two helium pressurization tanks to achieve optimal engine feedline pressures; pressure- control systems; a sensing and command system with drive electronics; propellant lines with isolation valves, filters, and additional ground test instrumentation. The facility was outfitted with a diffuser to enable pump down action to create a low pressure on the exit plane of the OMS engine as it operated, as testing was conducted in ambient pressure conditions.

In a test stand at White Sands, NASA and ESA used the PQM to complete 48 hot firing tests and 3 discrete pressurization tests conducted in two phases and concluding on Jan. 21, 2020. The data obtained during this test campaign was critical to the certification of the Orion/European Service Module propulsion system for
the Artemis I and II missions and beyond. The focus on and the propulsion subsystem, as well as the performance of the pressurization control assembly. Five additional hot fire tests were conducted utilizing the auxiliary
engines on PQM from Sept. 29 to Nov. 4, 2020. There were also isothermal and gradient tests conducted on the pressure control assembly valves in 2020 that involved chilling the valves and performing multiple cycles to see if anomalies could be re-created. The first gradient test was performed April 4 and repeated through April 21 of that year. A second gradient test was performed on May 5 and isothermal testing started May 12 and concluded June 3.

The first phase of propulsion testing had no active pressure regulation of the propellant tanks, with testing conducted in a "blow down" mode in which the tanks were pressurized prior to the start of each test. The second phase included active pressure regulation. The testing was divided into different campaigns and operated various combinations of the engines to simulate flight test scenarios, evaluate for engine cross talk or fluidic hammer in the propulsion system, evaluate engine performance, and test the pressurization cross feed assembly. Some campaigns evaluated system performance with saturated versus unsaturated propellant, or the amount of helium dissolved in the fuel and the oxidizer.

During testing, teams also successfully simulated the most taxing situation the spacecraft's engines could encounter after launch. In one of the high-profile tests, engineers performed a 12-minute propulsion test simulating an abort- to-orbit scenario, in which the spacecraft's service module must place Orion in a safe orbit because of a problem after the abort system has been jettisoned. The test, which was the longest ever continuous burn conducted on an OMS engine to date at White Sands, used the PQM to fire the Orbital Maneuvering System engine, eight auxiliary thrusters and six reaction control thrusters.

Over the course of this two-and-a-half-year test campaign over 15,000 kg of MON (mixed oxides of nitrogen – or propellant-grade nitrogen tetroxide) and almost 9,000 kg of MMH (monomethylhydrazine) hypergolic propellant were consumed.

PQM height, including frame	4.6 m
Base dimensions	4.9m x 4.872m
PQM mass (empty)	12,000 kg
PQM mass (filled)	22,000 kg

PQM Testing

Date	Test	Length	OMS-E	AUX	RCS
4/2/19	RCS Acceptance	11:49	N	N	Y
4/4/19	Waterhammer & Crosstalk Seq 1	12:32	Y (52 sec)	Y	Y
4/8/19	Waterhammer & Crosstalk Seq 2	13:57	N	Y	Y
4/11/19	Waterhammer & Crosstalk Seq 3	2:30	N	Y	Y
4/15/19	GNC Flight Profiles (ATO Light)	21:15	Y (15 sec)	Y	Y
4/29/19	OMS-E Feed System Test Seq C, Run 1ab	11:17	Y (5 sec)	N	N
4/29/19	Waterhammer & Crosstalk Seq 1	12:32	Y (52 sec)	Y	Y
5/2/19	OMS-E Feed System Test Seq C, Run 2	11:17	Y (5 sec)	N	N
5/2/19	RCS & Aux	5:10	N	Y	Y
5/6/19	OMS-E Feed System Test Seq C, Run 4	11:17	Y (5 sec)	N	N
6/20/19	Waterhammer & Crosstalk Seq 3	2:30	N	Y	Y
6/20/19	OMS-E Delta Qual Seq B, Run 2	11:32	Y (20 sec)	N	N
7/15/20	Waterhammer & Crosstalk Seq 2	13:57	N	Y	Y
7/17/19	GNC Flight Profiles (ATO Light)	21:16	Y (15 sec)	Y	Y
7/19/19	OMS-E Delta Qual Seq B, Run 1	11:32	Y (20 sec)	N	N
7/19/19	OMS-E Feed System Test Seq C, Run 3	11:17	Y (5 sec)	N	N
8/5/19	Full ATO Operation	23:21	Y (11.5 min)	Y	Y
9/9/19	Long duration Aux (TEI)	1:35:00	N	Y	Y
9/25/19	Waterhammer & Crosstalk Seq 4	13:57	N	Y	Y
12/5/19	Cross-Feed Test (BOL, highest Aux regulation frequency)	1:50	N	Y (70 sec)	Y
12/5/19, 12/10/19	Cross-Feed Test (BOL, highest OMS-E regulation frequency) (Runs 1&2 on 12/5, Run 3 on 12/10)	1:40	Y (70 sec)	N	Y
12/10/19	Cross-Feed Test (BOL, highest OMS-E regulation frequency) (Runs 1&2 on 12/5, Run 3 on 12/10)	1:40	Y (70 sec)	N	Y
12/13/19	Cross-Feed Test (EOL, lowest regulation frequency)	9:20	Y (530 sec)	N	Y
1/14/20	Aux Passenger Test 1 (includes 1.65 MR test for EM III+)	8:48	N	Y	N
1/21/20	Aux Passenger Test 2	5:05	N	Y	N

▲   An Orion boilerplate test article (BTA) drop test takes place at the Hydro Impact Basin at the Landing and Impact Research Facility at NASA's Langley Research Center on July 21, 2011.

Langley Research Center

NASA's Langley Research Center is the nation's first civilian aeronautics laboratory and the agency's original field center. Researchers at Langley work to make revolutionary improvements to aviation, expand understanding of Earth's atmosphere, and develop technology for space exploration. Langley sits on 764 acres in Hampton, Virginia, and employs about 3,400 civil servants and contractors.

Water impact testing is performed on Orion at the Hydro Impact Basin at the Landing and Impact Research Facility at Langley. The water basin is 115 feet long, 90 feet wide and 20 feet deep. The basin is located at the west end of the center's historic gantry. The gantry is a 240-feet high, 400-feet long, 265-foot-wide A-frame steel structure, built in 1963 and is where Neil Armstrong trained to walk on the Moon. The gantry provides the ability to control the orientation of the test article while imparting a vertical and horizontal impact velocity, which is required for human rating vehicles.

Water Impact Testing

When astronauts return to Earth in the Orion spacecraft, they will enter on an extremely hot and fast journey through the atmosphere before splashing down in the Pacific Ocean. To protect the crew on landing, water- impact testing evaluates how the spacecraft may behave in parachute-assisted landings in different wind conditions and wave heights.

The Structural Passive Landing Attenuation for Survivability of Human Crew (SPLASH) project team at NASA Langley conducted water-impact testing to provide high-fidelity data of the forces that the Orion spacecraft structure and its astronaut crew would experience during landing, helping to protect the crew and informing future designs.

Boilerplate Test Article (BTA)

Engineers conducted three vertical drop tests and six swing drop tests of an Orion low-fidelity mock-up, called the Boilerplate Test Article (BTA), between July 2011 and January 2012. The BTA tests were used to shakedown the test procedures and to provide more than 160 channels of data to assist engineers in the development of analytical models. Then, another ten vertical drops of the BTA were conducted in 2012 to measure the rate at which the loads travel through the structure and help fine-tune the way NASA predicts Orion's landing loads.

Ground Test Article (GTA)

After BTA, the higher structural fidelity Ground Test Article (GTA), with the heat shield from the Exploration Flight Test- 1 vehicle, was water impact tested between April 2016 and September 2016. The GTA tests included five vertical tests and five swing tests. There were 608 sensors on-board the GTA, compared to 155 sensors for the BTA. Over 500 channels of data were collected for the GTA including data on two anthropomorphic test devices (ATDs), or automotive test dummies, since they are the closest representation of the human body.

▼   The fourth and final water impact test of the Orion Structural Test Article was successfully completed at NASA's Langley Research Center in Hampton, Virginia on April 13, 2021.

The ATDs were of a 5th percentile female (weighing 119 pounds) and a 95th percentile male (weighing 236 pounds), representing the lower and upper range of astronauts. The data taken from the ATDs was used to assess safety of the crew and to calibrate and improve analytical models of the Orion vehicle. The campaign of swing and vertical drops simulated water landing scenarios to account for different velocities, parachute deployments, entry angles, wave heights and wind conditions the spacecraft may encounter when landing in the ocean.

Structural Test Article

Based on the final design for the configuration that will fly on Artemis II, NASA's first mission with crew, the Structural Test Article (STA) built by Lockheed Martin underwent water-impact testing at Langley in March-May 2021. The tests focused on the extreme limits of Orion's structural "twin" and less about reducing model uncertainty, as the BTA and GTA tests were. These tests looked similar to the previous, however, the STA crew module is more flight-like in its configuration, with several structural updates and modifications based on data from wind tunnel tests and Exploration Flight Test-1. Three vertical water impact tests and one swing water impact test were conducted in this series. Data from the drop tests will be used for final computer modeling for loads and structures prior to Artemis II.

▲ On Dec. 2, 2020, researchers at NASA's Langley Research Center unwrap Orion's structural test article to be used in a series of water impact tests that will prepare the final computer modeling for loads and structures prior to Artemis II.

Boilerplate Test Article (BTA) Phase 0 and 1 Testing	
Date	**Test Configuration**
7/12/11	Phase 0 Test #1 (Vertical, rigid heat shield) 38.4° pitch, -0.4° roll, Stable 1
7/21/11	Phase 0 Test #2 (Vertical, rigid heat shield) 20.3° pitch, -0.5° roll, Stable 1
8/3/11	Phase 0 Test #3 (Vertical, rigid heat shield) 45.8° pitch, -1.0° roll, Stable 2
10/18/11	Phase 1 Test #1 (Swing, flexible heat shield) 40.2° pitch, 31° roll, Stable 2
10/27/11	Phase 1 Test #2 (Swing, flexible heat shield) 15° pitch, 30.5° roll, Stable 1
11/8/11	Phase 1 Test #3 (Swing, flexible heat shield) 15.8° pitch, 0.3° roll, Stable 2
12/1/11	Phase 1 Test #4 (Swing, flexible heat shield) 23.4° pitch, 179.6° roll, Stable 1
12/13/11	Phase 1 Test #5 (Swing, flexible heat shield) 26° pitch, -1.0° roll, Stable 1
1/6/12	Phase 1 Test #6 (Swing, flexible heat shield) 46.1° pitch, -1.1° roll, Stable 2
9/20/12	Phase 1 Test #7 (Vertical) 45.5° pitch, -1.8° roll, Stable 2

Boilerplate Test Article (BTA) Vertical Drop Testing

Date	Test Configuration
8/23/12	Test #1 (Vertical) 2.24 ft., 16.4° pitch, -1.6° roll, Stable 1
8/27/12	Test #2 (Vertical) 2.24 ft., 16°pitch, -1.6° roll, Stable 1
8/28/12	Test #3 (Vertical) 2.24 ft., 16.4° pitch, 5.4° roll, Stable 1
8/29/12	Test #4 (Vertical) 8.96 ft., 27.1° pitch, -0.8° roll, Stable 1
8/30/12	Test #5 (Vertical) 8.96 ft., 28.1° pitch, 1.2° roll, Stable 1
8/30/12	Test #6 (Vertical) 8.96 ft., 27.7° pitch, -4.5° roll, Stable 1
9/7/12	Test #7 (Vertical) 8.96 ft., 27.4° pitch, -1.6° roll, Stable 1
9/12/12	Test #8 (Vertical) 24 ft. 10.75 in., 42.2° pitch, -0.2° roll, Stable 1
9/13/12	Test #9 (Vertical) 24 ft. 10.75 in., 42.2° pitch, 0° roll, Stable 1
9/13/12	Test #10 (Vertical) 24 ft. 10.75 in., 42.3° pitch, -1.0° roll, Stable 1
9/22/12	Test #11 (Vertical) 24 ft. 10.75 in., 41.8° pitch, -0.8° roll, Stable 1

Ground Test Article (GTA) Testing

Date	Test Configuration
4/6/16	Test #1 (Vertical) -34° pitch, 0° roll, Stable 1
4/13/16	Test #2 (Vertical) -45° pitch, 0° roll, Stable 1
4/19/16	Test #3 (Vertical) -23° pitch, 0° roll, Stable 1
5/11/16	Test #4 (Vertical) -23° pitch, 0° roll, Stable 1
6/8/16	Test #5 (Swing) +26° pitch, 180° roll, Stable 1
6/24/16	Test #6 (Swing) -20.3° pitch, 0° roll, Stable 1
7/21/16	Test #7 (Swing) -50.2° pitch, 0° roll, Stable 2
8/4/16	Test #8 (Swing) -43.8° pitch, 30° roll, Stable 2
8/25/16	Test #9 (Swing) -26° pitch, 90° roll, Stable 1
9/7/16	Test #10 (Vertical) -15° pitch, 0° roll, Stable 1

Structural Test Article (STA) Testing

Date		Test Configuration
2/11/2021	2/22/2021	Final waterproofing
2/22/2021		STA ready for WIT
2/23/2021		Weight & CG Test/TRR STA WIT/Clear RFAs
3/4/2021		Test #1 (Vertical) 21° pitch, 0° roll, Stable 1
3/11/2021		Test #2 (Vertical) 21° pitch, 0° roll, Stable 1
3/18/2021		Test #3 (Vertical) 21° pitch, 0° roll, Stable 1
3/23/2021	4 /12/2021	STA WIT pre-swing test activities
4/13/2021		Test #4 (Swing) 43°-46.5° pitch, 0° roll, Stable 2
4/23/2021	5 /6/2021	STA WIT close out activities/prepare/store to ship

NASA's Michoud Assembly Facility in New Orleans is an 832-acre site managed for NASA by the Marshall Space Flight Center in Huntsville, Alabama.

Michoud Assembly Facility

For more than half a century, NASA's Michoud Assembly Facility in New Orleans has been the agency's rocket factory and the nation's premiere site for manufacturing and assembly of large-scale space structures and systems. The government-owned manufacturing facility is one of the largest in the world, with 43 acres of manufacturing space under one roof. Michoud is managed by NASA's Marshall Space Flight Center in Huntsville, Alabama, with several areas of the facility used by commercial firms or NASA contractors. Manufacturing and assembly of some of the largest parts of NASA's Space Launch System and Orion spacecraft take place at this facility.

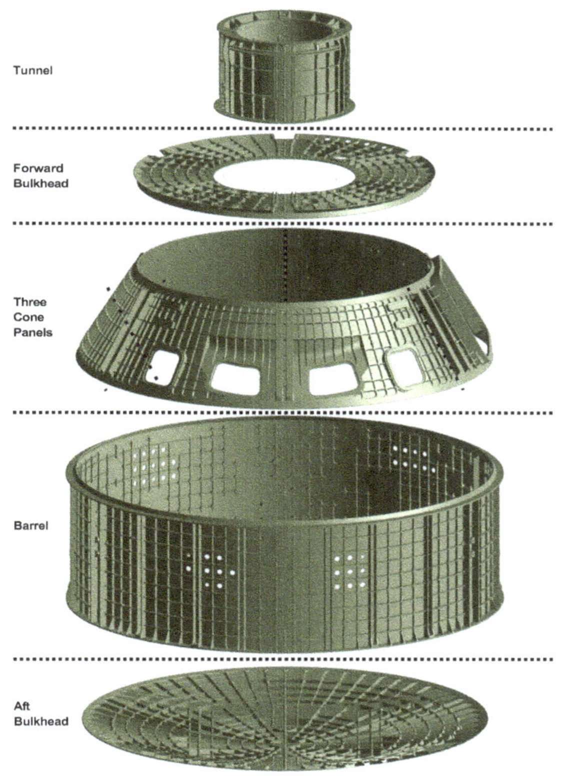

Seven Welds of Orion's Pressure Vessel	
Pressure Vessel Cone	
PVC-1	D-E (hatch) to A-F (windows) cone weld
PVC-2	A-F to B-C cone weld
PVC-3	B-C to D-E cone weld
Pressure Vessel	
PV-1	Tunnel to forward bulkhead weld
PV-2	Forward bulkhead to top of cone panels weld (becomes the forward structure)
PV-4	Barrel to aft bulkhead weld (becomes the aft structure)
PV-3	Closeout weld - forward structure to aft structure weld (bottom of cone panels to barrel)

Crew Module Pressure Vessel

Orion's crew module pressure vessel, the underlying frame of the crew module that provides an air-tight, habitable space for astronauts during the mission, is assembled at Michoud by prime contractor Lockheed Martin. The pressure vessel is designed to withstand the harsh and demanding environment of deep space and is the core structure upon which all the other elements of Orion's crew module are integrated. The pressure vessel's exterior is comprised of seven large aluminum alloy elements that are manufactured by two key subcontractors, AMRO Fabricating Corp. in South El Monte, CA and Ingersoll Machine Tools in Rockford, IL. AMRO provides the pressure vessel's three cone panels and Ingersoll provides the barrel, tunnel, forward and aft bulkheads.

Once all the elements arrive at Michoud, they are welded together in detailed fashion. The cone panels are joined with three welds to form the angled mid-section of the pressure vessel where the windows and hatch are located. Another weld connects the tunnel and the forward bulkhead, which is at the top of the spacecraft and houses many of Orion's critical systems, such as the parachutes that deploy during entry into Earth's atmosphere. Orion's tunnel, with a docking hatch, will allow astronauts to move between the crew module and other spacecraft. The spacecraft's barrel section, which is the largest single piece of the spacecraft, is welded to the aft bulkhead, which is the bottom portion of the vehicle, to form the aft structure. Technicians install Orion's backbone inside the aft structure.

The backbone assembly is a bolted structure consisting of nine pieces that are assembled prior to being installed inside the pressure vessel. The forward bulkhead is welded to the top of the cone panels in what is called the forward structure weld. The seventh and final major structural weld, also called the closeout weld, combines the forward and aft structures to complete the pressure vessel. Once welding of the crew module's pressure vessel is complete, it is shipped to NASA's Kennedy Space Center where it undergoes further assembly.

The Orion crew module pressure vessel is joined using a method called friction-stir welding. Friction-stir welding produces incredibly strong bonds by transforming metals from a solid into a "plastic-like" state, and then using a rotating pin tool to soften, stir and forge a bond between two metal components to form a uniform welded joint, a vital requirement of next-generation space hardware.

The technique is also being used at Michoud to weld the massive barrels of the SLS core stage to make the more than 200-foot tall structure.

Engineers undertake a meticulous process to prepare for welding. They clean the segments, coat them with a protective chemical and prime them. They then outfit each element with strain gauges and wiring to monitor the metal during the fabrication process. Prior to beginning work on the pieces destined for space, technicians weld together a pathfinder, a full-scale version of the current spacecraft design, to refine their techniques and ensure proper tooling configurations.

The pressure vessel for Artemis I was completed at Michoud in January 2016, and the pressure vessel for Artemis II was completed in August 2018. The Artemis III pressure vessel was completed in September 2021.

▼   On Jan. 13, 2016, technicians at Michoud Assembly Facility in New Orleans finish welding together the primary structure of the Orion spacecraft destined for deep space on Artemis I.

Other Orion Components

Work on Orion structures at Michoud goes beyond assembling the pressure vessel. The team provides design, manufacturing, assembly, integration, and test capabilities for key elements of Orion.

Other work on the crew module includes fabricating 227 Avcoat billets (117 standard blocks, 57 thick blocks, 53 curved blocks) that will be used to make the heat shield. The wall panels for the crew module's environmental control and life support systems (ECLSS) are also laid up, assembled, and tested at this facility. The assembly and testing of the module's docking hatch also takes place here, as well as the assembly of the four crew seats.

Work on the service module at Michoud includes laying up, trimming, assembling and testing 22 panels (six aft, six forward, seven outboards, and three access covers) for the crew module adapter, which will connect the service module to the crew module. The trimming, assembly, and testing of three spacecraft adapter jettison panels, which

will protect the service module from the environment during ascent, also takes place here. The same is done for the spacecraft adapter cone, which connects to the bottom of the service module and will later join another adapter connected to the top of the rocket's interim cryogenic propulsion stage (ICPS).

Structural components of the launch abort system (LAS) are assembled at Michoud, including two fillet panels, the motor adapter truss assembly, which connects the crew module to the LAS, and four ogive panels. The fillet and ogive panels also undergo testing here. LAS tower components assembled at Michoud include the nose cone and aft and forward interstage, the latter of which is also tested. Components of the raceway, which runs the electronic communication lines through the system, are assembled here, including eight raceway covers, and 24 access covers. Eighteen aero seals and 12 tangential fittings are also assembled at Michoud.

The ogive panels for the Artemis II Orion launch abort system being assembled at Michoud Assembly Facility in New Orleans.

Engineers from Ames Research Center and Marshall Space Flight Center remove samples from the Orion heat shield that flew on Exploration Flight Test-1 in 2014. The heat shield protected the spacecraft from temperatures reaching 4,000 degrees Fahrenheit.

Marshall Space Flight Center

NASA's Marshall Space Flight Center in Huntsville, Alabama, plays a vital role in the Artemis program and has provided significant fabrication and test support to the Orion program since late 2012. Marshall is the lead center for the Space Launch System and manages NASA's Michoud Assembly Facility in New Orleans, where the core stage of the Space Launch System is manufactured.

Marshall also provides support to NASA's lunar Gateway; making critical contributions to the Human Landing System program and managing all science payloads and science communications on board the International Space Station.

Orion Work

More than 1,500 parts for the spacecraft have been fabricated in Marshall's machine shop. The largest machining effort was the dissection for analysis of the Exploration Flight Test-1 heat shield after its return to Earth. The insulation materials contained in the heat shield protect the crew during the return to Earth. State- of-the-art machinery at Marshall was used to remove all the material for analysis, thereby confirming analytical models used to estimate insulation design and thickness for future flights.

Among the many component parts machined were clips, sleeves and rod ends used for unique connections on the Orion spacecraft.

Marshall also resolved processing questions on the Orion effort related to additive manufacturing. The field center's advanced additive manufacturing laboratory helped to define a process to successfully fabricate complicated parts and shared the process solution with commercial additive manufacturing shops to aid in 3D-printed parts fabrication for Orion.

The center also provides insight and oversight of environmental control and life support systems (ECLSS) for Orion's Crew and Service Module Office. Marshall assists Orion's European Service Module Integration Office by providing Orion isolation valve refurbishment and Orion gas valve seat material testing, as well as support for European Service Module propulsion systems with subsystem-level hot fire test operations, data review, verification-related review, and closure reports and activities.

Other contributions include natural environments support related to space and terrestrial environments including radiation, plasma, sea states, atmosphere, and winds; and plume and aerothermodynamic analysis and consultation as requested by Orion's Vehicle Integration Office.

Marshall also has performed various test series for Orion including vibration, thermal-vacuum, pyro-shock, tensile, and saltwater immersion testing. Numerous Orion parts, including the windows, were subjected to space-like environmental testing in Marshall's environmental test lab. This testing confirmed the ability of the hardware to withstand the harsh environment of space.

The launch abort system (LAS) integration and transition to production operations is managed by Marshall with Lockheed Martin as the prime contractor. Marshall also provides fabrication, production and assembly support to NASA teams across the Orion Program, including the Orion Production Operations Office. This effort includes integration of Orion and the SLS rocket.

▲ NASA's Orion spacecraft is prepared for its final set of environmental tests at NASA Glenn Research Center's Neil A. Armstrong Test Facility in Ohio on Feb. 21, 2020.

Neil A. Armstrong Test Facility

Neil A. Armstrong Test Facility, formerly known as Plum Brook Station, is a remote test facility for NASA's Glenn Research Center in Cleveland, Ohio. Located on 6,400 acres in the Lake Erie community of Sandusky, Neil A. Armstrong Test Facility is home to four world-class test facilities, which perform complex and innovative ground tests for the international space community.

The Space Environments Complex (SEC) at Armstrong Test Facility houses the world's largest and most capable space environment simulation facilities. This includes the Space Simulation Chamber, which simulates the thermal and vacuum conditions of space, and provides an environment to test electromagnetic interference; the Reverberant Test Facility, a 100,000 cubic foot reverberant acoustic chamber; and the Mechanical Vibration Facility, a vibration table capable of testing an entire spacecraft in all three axes. Major Orion-related tests have been conducted at the SEC, including the simulated space environments testing of the Artemis I spacecraft.

Environmental Testing

The Orion program has leveraged the unique facilities at Armstrong Test Facility to simulate environments the spacecraft will experience through launch, travel in deep space, and recovery, as well as evaluate the spacecraft's structure and systems in those conditions.

An international team of engineers and technicians completed three months of rigorous simulated in-space environments testing of the integrated crew and service module from December 2019 to mid-March 2020 for Artemis I. This testing was completed in two phases inside the Space Simulation Chamber.

Integrated Spacecraft Thermal Balance (TBAL) / Thermal Vacuum (TVAC) Testing

The first phase of testing was a thermal vacuum test lasting 47 days, while Orion's systems were powered-on under vacuum conditions that simulate the space environment. The spacecraft was subjected to extreme temperatures, ranging from -250 to 300-degrees Fahrenheit, to replicate flying in-and-out of sunlight and shadow in space. To simulate these conditions, a specially designed piece of hardware, known as the Heat Flux System, was used to heat specific parts of the spacecraft at any given time. It was also surrounded on all sides by a set of large panels called a cryoshroud that provided the cold background temperatures of space. The testing ensured Orion's systems perform correctly in extreme flight conditions.

Integrated Spacecraft Electromagnetic Interference/Compatibility (EMI/EMC) Testing

The second phase of testing was a 14-day electromagnetic interference and compatibility test. All electronic components have an electromagnetic field that can affect other electronics nearby. This testing ensured the spacecraft's electronics work properly when operated at the same time, as well as when bombarded by external sources. The test campaign confirmed the spacecraft's systems perform as designed, ensuring safe operation for the crew during future Artemis missions.

◀ Glenn Research Center's Neil A. Armstrong Test Facility (formerly Plum Brook Station) in Ohio houses the world's largest space simulation vacuum chamber where the Orion spacecraft, shown here on March 11, 2020, was rigorously tested for Artemis missions to the Moon.

Neil A. Armstrong Test Facility Fun Facts

Space Power Facility

» The Space Power Facility, or SPF, inside NASA's Neil A. Armstrong Test Facility in Sandusky, Ohio,
houses the world's largest and most powerful space environment simulation facilities.

» The opening sequences of the movie The Avengers were filmed at the SPF.

» Though it was built in the 1960s to test nuclear propulsion systems, the SPF was never used for nuclear testing.

Space Simulation Vacuum Chamber

» The Space Simulation Vacuum Chamber is the world's largest, measuring 100 feet in diameter by 122 feet high.

» The doors of the test chamber are 50 by 50 feet.

» The chamber floor was designed for a load of 300 tons.

» It is enclosed in a concrete chamber that varies in thickness from 6 to 8 feet.

» The vacuum chamber can simultaneously replicate the heat and cold of deep space.

» The vacuum chamber, at ambient pressure, contains 30 tons of air (the equivalent of two large elephants). When pumped down, the weight is equivalent to one soda can pull-tab.

» The vacuum chamber is made of the same amount of aluminum required to fabricate 1 billion soda cans.

» The vacuum chamber was used to test the airbag landing systems for the Mars Pathfinder and the Mars Exploration Rovers, Spirit and Opportunity, under simulated Mars atmospheric conditions. It also was used to test critical components of the International Space Station's electrical power system.

Reverberant Acoustic Test Facility

» The Reverberant Acoustic Test Facility (RATF) is the world's most powerful spacecraft acoustic test chamber. There are 36 concert-sized speakers on the wall.

» The RATF can simulate the noise of a spacecraft launch up to 163 decibels or as loud as the thrust of 20 jet engines.

» The Orion spacecraft is tested to withstand the acoustic forces of a launch or mission abort by
blasting the ground test vehicle with more than 150 decibels of sound, which is what you would feel and hear if you stood 50 yards from a jet aircraft engine.

» The door to the RATF weighs 675,000 pounds.

Mechanical Vibration Facility

» The Mechanical Vibration Facility (MVF) is the world's highest capacity and most powerful spacecraft shaker system.

» The 4.5-million-pound concrete MVF seismic mass is anchored to bedrock using 104 tension anchors, the longest of which are 65 feet long.

» The vibration-simulation table is 22 feet wide and weighs 55,000-pounds. It simulates the shaking a spacecraft could expect when launching on top of a rocket.

Operations and Checkout Facility, Kennedy Space Center

The Neil Armstrong Operations & Checkout (O&C) Building at Kennedy Space Center in Cape Canaveral, Florida, has played a vital role in NASA's spaceflight history. Built in 1964, the facility was used during the Apollo program to process and test the command, service, and lunar modules. The facility is being used today to process and assemble the Orion spacecraft. Orion is tested, assembled, and readied in a large room in the O&C called a high bay, operated by Lockheed Martin, that functions as a high-tech factory. The high bay includes unique tooling stations, test fixtures, chambers, and clean rooms for the buildup and testing of Orion.

Environmental Testing

Environmental testing, including thermal cycle testing and acoustic testing, takes place in the high bay of the O&C. Environmental testing is performed to simulate environments the spacecraft will experience through launch, travel in deep space, and recovery, as well as evaluate the spacecraft's structure and systems in those conditions.

▼ The Orion crew module for Artemis I is blasted with 141 decibels of extreme vibrations during acoustics testing in the Neil Armstrong Operations and Checkout Building high bay at NASA's Kennedy Space Center in Florida, simulating what the vehicle will experience in space.

▲   Technicians assist as the Orion crew module for Artemis I is moved toward the thermal chamber in the Neil Armstrong Operations and Checkout Building high bay at NASA's Kennedy Space Center in Florida.

Crew and Service Module Direct Field Acoustic Testing

The crew module and service module undergo direct field acoustic testing separately to ensure they are prepared to endure the noises of launch and ascent to space. During this testing, the crew or service module is surrounded with speakers and exposed to maximum acoustic levels that the vehicle will experience in space. Engineers secure the module inside a test cell and then attach microphones, strain gauges, and accelerometers to it. The module is blasted with extreme vibrations and acoustic levels up to 141 decibels – as loud as a jet engine during takeoff – to ensure the spacecraft and its systems can withstand the noise expected during launch.

▲ Orion's service module for NASA's Artemis I mission completes acoustic testing inside the Operations and Checkout Building at NASA's Kennedy Space Center in Florida on May 25, 2019.

Crew Module and Service Module Thermal Cycle Testing

Thermal cycle testing is performed separately on the crew and service module inside a specially constructed thermal cycle chamber. Over several days, the crew or service module is rapidly cycled between hot and cold temperatures to thermally stress the hardware and ensure the workmanship of the module's critical hardware and its subsystem operations.

Engineers successfully completed a thermal cycle test on the Artemis I crew module in February 2018. The cycle of temperatures for the initial thermal test ranged from 29 to 129 degrees Fahrenheit during 105 hours of testing. Engineers successfully completed a thermal cycle test on the Artemis I service module in December 2019.

Acceptance and Workmanship Testing

The Orion spacecraft undergoes assembly and testing in the high bay of the O&C Building to ensure its workmanship before ground processing and integration with the SLS rocket.

Proof Pressure Testing

Engineers conduct a series of tests on the spacecraft's pressure vessel, or underlying structure of the crew module. In a test stand inside the proof pressure cell, technicians attach hundreds of strain gauges to the interior and exterior surfaces of the vehicle. The strain gauges measure the strength of the welds on the spacecraft as it is pressurized at incremental steps over two days to reach the maximum pressure it is expected to encounter during flight.

▲ Lockheed Martin engineers and technicians prepare the Orion pressure vessel for a series of tests inside the proof pressure cell in the Neil Armstrong Operations and Checkout Building at NASA's Kennedy Space Center in Florida.

▲ The Orion crew module for Artemis I is secured in a workstation in the Neil Armstrong Operations and Checkout Building high bay at NASA's Kennedy Space Center in Florida.

The tests confirm that the weld points of the underlying structure will contain and protect astronauts during the launch, in-space, entry into Earth's atmosphere, and landing phases on Artemis missions.

Proof pressure testing was completed on the Artemis I pressure vessel in May 2016.

Crew Module and Service Module Functional and Performance Testing

Orion crew modules undergo initial power-on events, which are the first time the vehicle management computers and the power and data units are installed on the crew module, loaded with flight software and tested. These tests verify the health and status of Orion's core computers and power and data units, and ensure the systems communicate precisely with one another to accurately route power and functional commands throughout the spacecraft.

Orion service modules also separately undergo initial power-on events. The initial power-on tests allow technicians to check all cables are connected and data is being transferred at the speeds required by Orion, as well as the power distribution across the module.

In August 2017, the Artemis I crew module was powered on for the first time. The Artemis I service module, integrated with the crew module adapter, was powered on for the first time in February 2019.

After initial power-on, the Orion crew module and service module undergo separate functional testing, which ensures each of the module's systems powers on and functions as designed.

After functional testing is completed, separate performance testing of the crew module and service module takes place, which verifies that each module's systems not only power on but are functioning at the correct parameters. Performance testing also takes place after the two modules are mated.

Spacecraft Integration and Final Testing

After functional and performance testing takes place on the crew and service modules individually, they are moved to the Final Assembly and System Testing (FAST) cell, where they are integrated and put through their final system tests prior to rolling out of the O&C for commodity servicing and integration with the rocket.

Inside the FAST cell, the integrated spacecraft goes through a final round of testing and assembly, including end-to- end performance verification of the vehicle's subsystems, checking for leaks in the spacecraft's propulsion systems, installing its solar array wings, performing spacecraft closeouts and pressurizing a subset of its tanks in preparation for flight prior to being transported to the Multi-Payload Processing Facility (MPPF). From July to November 2019, engineers at Kennedy mated the Artemis I crew and service module inside the FAST cell, where the integrated spacecraft underwent functional and performance testing.

The Orion crew and service module stack for Artemis I is lifted out of the Final Assembly and Test (FAST) cell in the Neil Armstrong Operations and Checkout Building at NASA's Kennedy Space Center in Florida on Nov. 11, 2019.

The Artemis I spacecraft was then transferred to NASA's Armstrong Test Facility in Sandusky, Ohio, where it underwent full environmental testing from December 2019 to March 2020.

The vehicle then returned to the FAST cell in March 2020, to go through a final round of testing and assembly prior to being transported to the MPPF in January 2021.

Lockheed Martin Facilities

Development, testing, and assembly of Orion takes place at several different Lockheed Martin facilities across the country. This includes Lockheed Martin's Waterton Canyon campus near Denver, which has been a hub of space innovation since the 1950s and has more than 4,000 employees and a wide range of industry-leading design, manufacturing, and test facilities on site. Orion avionics and software testing also takes place at Lockheed's Exploration Development Laboratory located in Houston.

Lockheed Martin assembles Orion (Artemis I & II) at NASA's Michoud Assembly Facility and NASA's Kennedy Space Center in the Neil Armstrong Operations and Checkout Building, and will soon be utilizing the new STAR Center, or Spacecraft, Test, Assembly and Resource Center, in Titusville, Florida, to build and test large Orion elements for Artemis III and beyond.

▼ Engineers complete testing on a duplicate of Orion called the Structural Test Article (STA), needed to verify the spacecraft is ready for Artemis I. NASA and its prime contractor, Lockheed Martin, built the STA to be structurally identical to Orion's main spacecraft elements: the crew module, service module and launch abort system.

Structural Test Article

In June 2020, engineers completed testing on a duplicate of Orion called the Structural Test Article (STA), needed to verify the spacecraft is ready for its Artemis missions. NASA and its prime contractor, Lockheed Martin, built the STA to be structurally identical to Orion's main spacecraft elements: the crew module, service module and launch abort system, but missing critical non-structural items such as the vehicle's computers, propulsion, and seats not required for these tests.

Testing deliberately took the vehicle's structure to the edge of its design, simulating the harsh environments that physically affect the structures of the Orion spacecraft. Invaluable to the mission, the tests on Orion's "structural twin" at Lockheed Martin's facility in Colorado provide a way to validate Orion structurally and enable engineers to push the structure past design standards.

The STA testing required to qualify Orion's design began in early 2017 and involved 21 tests, using six different configurations — from a single element to the entire full stack — and various combinations in between. The different configurations simulated the different flight conditions, such as launch, return to Earth, parachute deployment, water landing, etc. Orion will go through during a mission. Teams worked round-the-clock for days at a time to prepare the tests, execute, tear down then reconfigure the STA for the next test, culminating in 330 actual days of testing. During some test phases, engineers pushed expected pressures, mechanical loads, vibration, and shock conditions up to 40 percent beyond the most severe conditions anticipated during the mission, analyzing data to confirm the spacecraft structures can withstand the extreme environments of space.

At completion, the testing verified Orion's structural durability for all flight phases of Artemis missions.

Structural Test Article (STA) Testing

Start	End	Config	Test	Component
4/11/17	4/14/17	PP	Proof Pressure	CM
5/27/17	9/8/17	12A	Stiffness/Qual	CM
7/10/17	8/21/17	6	Stiffness/Qual	LAS
8/23/17	9/1/17	10	Modal	ESM
1/15/18	1/21/18	9	Modal	CM/ESM
1/26/18	1/30/18	3	Modal	LAS/CM/ESM
6/11/18	6/14/18	13	Acoustic	LAS/CM/ESM
6/21/18	7/11/18	4	Modal	LAS/CM/ESM
7/24/18	7/24/18	15	Pyro Shock	LAS/CM R&R
8/14/18	8/15/18	18A	Pressure	CM
10/29/18	11/16/18	5	Stiffness	LAS/CM/ESM
1/15/19	1/31/19	18	Acoustic	CM (FBC GAP)
5/1/19	5/1/19	19	Jettison	CM (FBC)
6/3/19	9/13/19	7	Stiffness/Qual	ESM
6/8/19	7/3/19	20	Pyro Shock	CM (Mortar/Riser)
9/11/19	12/4/19	12	Stiffness/Qual	CM
1/22/20	1/29/20	11A	Modal	LAS/CM/ESM
3/26/20	4/13/20	14	Lightning	LAS/CM/ESM
6/1/20	6/1/20	16	Jettison	CM/ESM (SAJ)
6/17/20	6/17/20	17	Pyro Shock	CM/ESM R&R
7/31/20	10/9/20	6A	Stiffness/Qual & Modal	LAS
7/15/20	7/28/20	21	Pyro Shock	CM (DMJS)

Pressure Testing

Pushing and pulling with pressure that equates to 140 percent of the maximum expected loads during missions ensures the spacecraft structures can withstand intense loads at launch and entry into Earth's atmosphere.

Modal Testing

During modal testing, dynamic loads of pressure are applied to the spacecraft structures. With more than 20,000 parts making up Orion's service module alone, modal tests are needed to evaluate how all of the spacecraft components hold up to vibration, especially at connection points.

Stiffness/Qualification Testing

Stiffness/qualification testing applies pressure steadily and continuously to the spacecraft's structures. This tests how the structures will respond to the critical static loads it will experience on missions.

Acoustic Testing

Orion will need to withstand incredible force during its missions. Blasting the structures with sound waves simulates the vibrating rumble of launch, reaching more than 160 decibels.

Pyrotechnic Shock Testing

Shock tests recreate the powerful pyrotechnic blasts that are needed for critical separation events during flight, such as the launch abort system separating from the crew module after a successful launch.

Jettison Testing

Jettison tests mimic deployment mechanisms required to jettison the forward bay cover and to ensure components can endure the shock levels expected during flight.

Lightning Testing

Lightning tests evaluate potential flight hardware damage when the flight vehicle is exposed to a lightning strike prior to launch.

Integrated Test Lab

The Orion Integrated Test Lab (ITL) at Lockheed Martin's facility near Denver simulates the flight environment to test Orion's avionics and flight software functions for both Artemis I and II. NASA teams have done real-time monitoring and commanding of the ITL Orion vehicle in Denver from the Mission Control Center in Houston, during tests of the Orion communication system and simulations of Orion's uncrewed flight to the Moon for Artemis I. The ITL began integration for the Artemis I mission in January 2015.

Tests performed using the ITL are essential for identifying software problems and validating proper functionality and performance of the spacecraft avionics system. Orion's avionics and flight software functions tested at the ITL include ascent abort, safe mode, fault detection, isolation and recovery, optical navigation, maneuver plan management, and propulsion failure detection.

The ITL tests the millions of lines of software code with flight-representative engineering development unit avionics hardware in a flight-like configuration.

The lab runs full mission scenarios from pre-launch to landing, or specific phases and events, in real time. The lab has a fully integrated set of Orion's complex crew module and service module avionics, power, wiring, and guidance, navigation, and control hardware. The lab is the only one of its kind that can integrate hardware and software across all of Orion program elements, and the lab is also used by other programs such as Space Launch System, Exploration Ground Systems and the Flight Operations Division to conduct integrated Artemis mission testing.

An Integrated Test Lab rig configuration is the highest fidelity test platform that Orion avionics hardware and software would experience prior to actual testing regimens on the assembled vehicle. In this configuration, many actual vehicle avionics components are connected and loaded with associated flight software. Appropriate vehicle, environmental, and dynamics simulations, along with actual and/or simulated sensor/effectors and stimulators provide as close to a "test like you fly" environment as can be assembled within a lab setting.

The Integrated Test Laboratory (ITL) has a full set of avionics, harnessing, sensors, and flight software identical to that on Orion. Using this real flight-like hardware and software, the lab is used to simulate and test every aspect of the Artemis I mission from launch to splashdown.

Exploration Development Lab

The Exploration Development Laboratory in Houston (EDL-H) is a state-of- the-art test facility initially funded by Lockheed Martin and its teammates United Space Alliance and Honeywell. The EDL was designed to reduce cost and schedule risk by providing an early opportunity in the development phase of the program to perform systems- level avionics and software testing for Orion in a realistic environment. The EDL-H's location adjacent to NASA Johnson Space Center has enabled the Lockheed Martin team to work closely with NASA early in the development and testing phase to gain clarity on requirements, including early involvement and collaboration with astronaut flight crew members and flight controllers.

Initial testing of critical systems is done in the EDL-H, including the guidance, navigation, and control (GN&C), automated rendezvous and docking, crew interfaces, and software development processes. Avionics system testing was performed to reduce risk prior to Pad Abort-1 flight testing at White Sands Missile Range and Exploration Flight Test-1 at NASA's Kennedy Space Center. EDL testing also included system integration tests and mission tests that employ the team's "test like you fly" philosophy.

The EDL-H houses the Houston Orion Test Hardware, otherwise known as HOTH, which is used to perform early development, integration, dry-run testing of Orion avionics hardware and software, and associated internal and external crew module interfaces using flight- representative flight software and an appropriate suite of ground support tools, systems, and software.

STAR Center

The STAR Center, or Spacecraft, Test, Assembly and Resource Center, in Titusville, Florida, is a Lockheed Martin production center for Orion spacecraft that will fly on Artemis III and beyond. The center provides urgently needed space to expand and streamline manufacturing capacity.

Elements of the spacecraft that take large amounts of floor space and that are built and tested outside the normal spacecraft assembly flow will be moved to the STAR Center. This gives production teams more room at NASA's Operations & Checkout (O&C) Building to assemble and test more Orion spacecraft simultaneously and quicker.

The center's work will start with the Orion Artemis III spacecraft, with many components already in development. The STAR Center will provide assembly and test of Orion aeroshell heat shield and backshell panels, including thermal protection system installation; crew module and crew module adapter wire harness fabrication and testing; propulsion and environmental control and life support systems assembly and testing; and electrical ground support equipment production.

The facility features a class 7 clean room for spacecraft component assembly, a proof pressure cell for spaceflight readiness testing, and a 9,000-square-foot high bay with ceiling crane. It also includes nondestructive evaluation and inspection capability, which allows team members to test and analyze materials, parts, structures, or systems without damaging the original component, as well as a separate machine shop.

The STAR Center is also integrated into Lockheed Martin's Intelligent Factory Framework (IFF), an edge computing platform that secures, scales, and standardizes device connectivity through various IT platforms. This digital- first approach streamlines production and maximizes agility by connecting devices virtually. In addition, more than 30 machines at the STAR center will be connected to this IFF, as well as machines at NASA's O&C, giving all production team members at both facilities real-time access to valuable data. The center also employs remote access, monitoring and alerting technologies for equipment, plus smart tools such as virtual reality and augmented reality.

Other Facilities Testing

Launch Abort Motor Testing

The launch abort system motors are being rigorously tested before the first crewed launch of Orion on Artemis II. The abort motor, attitude control motor, and jettison motor have each completed final qualification testing. These tests look at the maximum high and low temperature conditions that a motor might see during a launch in Florida. They provide the data on how the motor reacts under hot or cold stressing conditions. All three motors were also flight tested during the Ascent Abort-2 demonstration in July 2019.

Abort Motor

The 17-foot-long, three-foot-diameter abort motor has a manifold with four exhaust nozzles and provides thrust to quickly pull the crew module to safety if problems develop during launch. The high-impulse motor is designed to burn most of the propellant within the first three seconds and burns three times faster than a typical motor of this size to immediately deliver the thrust needed to pull the crew module to safety. If needed during a launch mishap, the crew module would accelerate from zero to 400-500 mph in two seconds. The motor was built by Northrop Grumman and tested at its facilities in Promontory, Utah.

During a series of static-fire tests, the motor was fastened to a vertical test stand with its nozzles pointed toward the sky. Upon ignition, the abort motor fired for five seconds with the exhaust plume flames reaching up to 100 feet in height. As expected, the motor reached approximately 400,000 pounds of thrust in one-eighth of a second – enough thrust to lift 66 large SUVs off the ground. The tests verified the motor can fire within milliseconds when needed and will work as expected under extreme temperatures.

▲ Engineers successfully perform a ground firing static test of the abort motor for the Orion spacecraft launch abort system a at Northrop Grumman facility in Promontory, Utah, on June 15, 2017.

Weight	7,600 lbs. includes 4,500 lbs. propellant 400,000 lbs. peak thrust
Fuel Type	Solid Fuel - hydroxyl-terminated polybutadiene (HTPB)
Burn Duration	About 3 sec.

The first test, called **Qualification Motor-1 (QM-1)**, took place under high temperatures, with a target temperature of 100 degrees Fahrenheit. The test confirmed the system performed as intended under the maximum temperature conditions a motor might experience during a launch.

The second test, **Qualification Motor-2 (QM-2)**, took place under cold temperatures, with a target temperature of 27 degrees Fahrenheit. The test confirmed the system performed as intended under the minimum temperature conditions a motor might experience during a launch.

The third and final test, **Qualification Motor-3 (QM-3)**, took place under ambient conditions, or the conditions of the naturally surrounding environment. The test confirmed the system can perform as intended under average temperature conditions a motor might experience during a launch. The Orion program made a strategic decision to delay this test to incorporate new insulator material, called EPDM rubber. The high-performance rubber insulation is glued to the inside of the motor casing and the propellent is poured inside. Following this test, the LAS was officially qualified for flight with crew.

Abort Motor Testing in Promontory, UT

QM-1	June 15, 2017	Hot Condition	100°F
QM-2	Dec. 13, 2018	Cold Condition	27°F
QM-3	March 31, 2022	Ambient Condition (Motor with new insulation)	

▼ NASA, Northrop Grumman, and Lockheed Martin successfully perform a ground firing static test of the abort motor for the Orion spacecraft launch abort system at Northrop's facility in Promontory, Utah, on Dec. 13, 2018.

Attitude Control Motor

The attitude control motor steers the Orion crew module to safety and orients it with the heat shield facing the Earth in the case of an emergency. The motor is essential because it helps stabilize Orion and control its trajectory as it moves away from the rocket.

Built by Northrop Grumman and tested at their facility in Elkton, Maryland, the attitude control motor consists of a solid propellant gas generator and eight equally spaced valves capable of providing 7,000 pounds of thrust in any direction. The unique valve control system enables each valve to open and close, directing the flow of gas.

The attitude control motor has undergone three qualification tests. During each of the three 30-second hot-fire tests, the motor's eight high-pressure valves directed more than 7,000 pounds of thrust in multiple directions, proving the motor can provide enough force to orient Orion and its crew for a safe landing.

Weight	1,700 lbs.	
	includes 650 lbs. propellant	
	7,000 lbs. thrust	
Fuel Type	Solid Fuel - carboxyl-terminated polybutadiene (CTPB)	
Burn Duration	About 30 sec.	

▲ Engineers conduct a static hot-fire test of the Orion spacecraft's launch abort system attitude control motor at a Northrop Grumman facility on March 20, 2019, in Elkton, Maryland.

The first test, **Qualification Motor-1 (QM-1)**, took place under ambient temperature conditions, or the conditions of the naturally surrounding environment, at approximately 70 degrees Fahrenheit. The test confirmed the system performed as intended under average temperature conditions a motor might experience during a launch.

The second test, **Qualification Motor-2 (QM-2)**, took place under hot conditions at approximately 94 degrees Fahrenheit. The test confirmed the system performed as intended under the maximum temperature conditions this motor might experience during a launch.

The third test, **Qualification Motor-3 (QM-3)**, took place under cold conditions at approximately 30 degrees Fahrenheit. To demonstrate worst case conditions, the motor was ignited using one of two initiators and simulated high altitude vacuum conditions. The test confirmed the system performed as intended under the minimum temperature conditions a motor might experience during a launch.

Attitude Control Motor Testing in Elkton, MD

QM-1	March 20, 2019	Ambient Condition	70°F
QM-2	Aug. 22, 2019	Hot Condition	94°F
QM-3	Feb. 25, 2020	Cold Condition	30°F

Jettison Motor

In the event of an emergency, the jettison motor ignites to separate Orion's LAS structure from the spacecraft, which could then deploy its parachutes for a safe landing. During a nominal mission, the jettison motor activates after stage- two ignition to separate the LAS from the spacecraft as the crew members continue their journey. This critical task makes it the only motor on the LAS to fire on every mission. The motor was built by Aerojet Rocketdyne and tested by engineers at the U.S. Army Redstone Test Center on Redstone Arsenal in Huntsville, Alabama.

During each 1.5 second hot-fire test, the jettison motor successfully produced more than 40,000 pounds of thrust in different conditions it might encounter during launch.

The first test, called **Qualification Motor-1 (QM-1)**, took place under high temperatures, at 105.9 degrees Fahrenheit. The test confirmed the system performed as intended under the maximum temperature conditions a motor might experience during a launch.

The second test, **Qualification Motor-2 (QM-2)**, took place under cold temperatures, at 23.6 degrees Fahrenheit. The test confirmed the system performed as intended under the minimum temperature conditions a motor might experience during a launch.

The third test, **Qualification Motor-3 (QM-3)**, took place under ambient temperature conditions, or the conditions of the naturally surrounding environment, at 73.1 degrees Fahrenheit. The test confirmed the system performed as intended under average temperature conditions a motor might experience during a launch.

Weight	900 lbs.
	includes 360 lbs. propellant
	40,000 lbs. thrust
Fuel Type	Solid Fuel - hydroxyl-terminated polybutadiene (HTPB)
Burn Duration	About 1.5 sec.

▼ Engineers perform a hot-fire test of the jettison motor for the Orion spacecraft's launch abort system at the U.S. Army Redstone Test Center on Redstone Arsenal in Huntsville, Alabama on Feb. 25, 2020.

▲   Aerojet Rocketdynes completes a demonstration test on Aug. 28, 2019, with the first jettison motor for the Orion spacecraft's launch abort system   that utilized Aerojet's Orange, Virginia facility for propellant mixing and loading and motor final assembly.

Aerojet Rocketdyne also completed a demonstration test on Aug. 28, 2019, with the first jettison motor that utilized Aerojet Rocketdyne's Orange, Virginia, facility for propellant mixing and loading and motor final assembly. The Demonstration Motor-4, or DM-4 test, rounded out an extensive two-year transition effort that increased affordability and demonstrated that there was no loss of fidelity or quality in the transfer of the program from California to Virginia.

Jettison Motor Testing in Redstone Arsenal, AL

QM-1	March 20, 2019	Ambient Condition	70°F
QM-2	August 22, 2019	Hot Condition	94°F
QM-3	Feb. 25, 2020	Cold Condition	30°F

Parachute Testing

Orion's parachute system is designed to ensure a safe landing for astronauts returning from deep space missions to Earth in the crew module at speeds exceeding 25,000 mph. While the Earth's atmosphere acting on Orion's heat shield will initially slow the spacecraft down to 325 mph, the parachutes are needed to get to a safe landing speed of 20 mph or less, making it a critical system.

The parachute system includes 11 parachutes that begin deploying at just under five miles in altitude. A series of cannon-like mortars fire to deploy three of the four parachute types, pyrotechnic riser cutters, and more than 13 miles of Kevlar lines attaching the top of the spacecraft to 36,000 square feet of parachute canopy material. Within 10 minutes of descent through Earth's atmosphere, everything must deploy and assemble itself in a precise sequence to slow Orion and its crew for splashdown in the ocean. The parachute system also must be able to keep the crew safe in several failure scenarios, such as mortar failures that prevent a single parachute to deploy, launch vehicle aborts, or other conditions that produce loads close to the maximum material capability. Each of Orion's main parachutes weigh 270 pounds and is packed to the density of oak wood to fit in the top part of the spacecraft, but once fully inflated, the three mains cover almost an entire football field.

The system underwent 17 developmental and eight qualification tests at the U.S. Army's Yuma Proving Grounds in Arizona. The Engineering Development Unit (EDU) test series ran from 2011 to 2016 and tested parachute system design data. During the development series, engineers tested different types of failure scenarios and extreme descent conditions to refine the design and ensure

Orion's parachutes will work in a variety of circumstances. Qualification airdrop and ground testing took place from 2016 to 2018. During the qualification testing, engineers evaluated the performance of the parachute system during normal landing sequences as well as several failure scenarios and a variety of potential aerodynamic conditions to ensure astronauts can return safely from deep space missions.

▲ Engineers complete the final test to qualify Orion's parachute system for flights with astronauts at the U.S. Army's Yuma Proving Ground in Arizona on Sept. 12, 2018.

The Parachute Compartment Drop Test Vehicle (PCDTV), an aerodynamically stable dart-shaped test vehicle, was used to provide high dynamic pressure test conditions. The Parachute Test Vehicle (PTV), a less stable, capsule-shaped test vehicle, was used more often and provided flight-like wake environments during tests.

While airdrop testing was a vital, and very visible, component to the development of Orion's parachutes, ground testing and analysis were equally important to ensure success. Airdrop testing is expensive, time- consuming, and cannot physically reach all possible spaceflight deployment conditions, but its data helped generate computer models of parachute performance and allowed the team to evaluate the parachutes in altitude and airspeed regimes that were unable to be thoroughly drop tested. Repeated simulation of the parachutes with varied parameters, called the Monte Carlo method, allowed the team to estimate the bounds of what parachute loads and performance should be expected throughout the life of the program.

Ground testing of material capabilities was coupled with the parachute simulations to determine how much structural margin exists in the system. This combination of ground tests, airdrop tests, and analysis qualified the system for Artemis flights with astronauts. Orion's parachute system consists of: Three forward bay cover parachutes (FBCP) used in conjunction with pyrotechnic linear thrusters to ensure separation of the forward bay cover (FBC), which protects Orion and its parachutes during the heat of entry into Earth's atmosphere. The FBCP are packed using a hydraulic press, with forces as high as 3,000 pounds.

Two drogue parachutes used to slow and stabilize the crew module during descent and establish proper conditions for main parachute deployment to follow. The drogues are mortar deployed from the crew module forward bay at 100 feet per second (68 mph) minimum muzzle velocity. The drogues are packed using a hydraulic press, with forces as high as 10,000 pounds.

Three pilot parachutes used to lift and deploy the main parachutes from the crew module forward bay. They are mortar deployed from the crew module forward bay at 112 feet per second (76 mph) minimum muzzle velocity. The pilots are packed using a hydraulic press for convenience but are much lower density and can be "hand packed" if required.

Three main parachutes used to slow the crew module for landing to a speed that ensures astronaut safety. The mains are packed using a hydraulic press, with forces as high as 50,000 pounds. They are autoclaved with a vacuum applied to the parachute at 190 degrees Fahrenheit for 48 hours to help "set" the packing and remove atmospheric moisture.

The parachute system was developed and tested by NASA and the agency's contractor partners. Parachutes are designed and fabricated by Airborne Systems in Santa Ana, California; the mortars are provided through Lockheed Martin by General Dynamics Ordinance & Tactical Systems, located in Seattle; and project management is performed by Jacobs Engineering's Engineering Science Contract Group in Houston.

Parachute Fun Facts

» The Earth's atmosphere slows Orion down from 25,000 mph to about 325 mph during entry, but parachutes will slow it down the rest of the way to about 17 mph for a gentle splashdown in the ocean.

» The parachute system has a total of 11 parachutes, and they all have to work carefully together.

» There are three 7 ft. forward bay cover parachutes that remove the forward bay cover, two 23 ft. drogue parachutes that slow and stabilize the crew module and three 11 ft. pilot parachutes that help to deploy the three 116 ft. main parachutes.

» The diameter of the three main parachutes combined would cover a football field from 10-yard line to 10-yard line.

» Each main parachute has 80 suspension lines. Each suspension line is rated to carry at least 1,500 lbs., which is strong enough to hold 6 adults with some margin to spare.

» The suspension lines on the three main parachutes combined are approximately 10 miles total length.

» The main parachutes are packed using a hydraulic press, using forces as high as 50,000 lbs. Their density is approximately 44 lbs. per cubic foot, which is roughly the same density as oak (and a much higher density than pine).

Parachute Quick Facts

Three Forward Bay Cover Parachutes (FBCP)

Diameter	7 ft.
Length	100 ft.
Weight	8 lbs. each
Material	All Kevlar materials
Deployment Altitude	26,500 ft.
Deployment Vehicle Speed	475 ft. per sec. (324 mph)
Density	Approx. 49 lbs. per cubic ft. (roughly the same as oak)
Final Packed Size	7.2" by 6.9" (0.16 cubic ft.) cylinder

Three Pilot Parachutes

Diameter	10 ft.
Length	70 ft.
Weight	9 lbs. each
Material	Kevlar and Nylon materials
Deployment Altitude	9,500 ft.
Deployment Vehicle Speed	190 ft. per sec. (130 mph)
Density	Approx. 35 lbs. per cubic ft. (roughly the same as pine)
Final Packed Size	6.8" by 13.3" (0.3 cubic ft.) cylinder

Two Drogue Parachutes

Diameter	23 ft.
Length	100 ft.
Weight	60 lbs. each
Material	Kevlar and Nylon materials
Deployment Altitude	25,000 ft.
Deployment Vehicle Speed	450 ft. per sec. (307 mph)
Density	Approx. 40 lbs. per cubic ft. (Roughly the same as oak)
Final Packed Size	16.5" by 16.2" (2 cubic ft.) cylinder

Three Main Parachutes

Diameter	116 ft.
Length	220 ft.
Weight	270 lbs. each
Material	Kevlar and Nylon materials
Deployment Altitude	9,000 ft.
Deployment Vehicle Speed	190 ft. per sec. (130 mph)
Density	Approx. 44 lbs. per cubic ft. (Roughly the same as oak)
Final Packed Size	Approx. 7 cubic ft. irregular shape to fit into the vehicle forward bay

Parachute Engineering Development Drop Testing

Number	Date	Vehicle	Alt.	Parachute # in Cluster				Primary Test Objective(s)
				FBC	Drogue	Pilot	Main	
CDT 3-1	09/21/11	PCDTV	25 kft	-	2	3	3	Nominal system
CDT 3-2	12/20/11	PCDTV	25 kft	-	2	2	2	Drogue skip 2nd, Pilot & Main fail to deploy
CDT 3-3	02/29/12	PTV	25 kft	-	2	3	3	Nominal system with flight-like wake behind PTV
CDT 3-4	04/17/12	PCDTV	25 kft	-	2	3	3	High Q Drogue deploy, Main skip 2nd
CDT 3-5	07/18/12	PTV	25 kft	-	2	3	3	Main skip 1st
CDT 3-6	08/28/12	PCDTV	25 kft	-	2	3	3	Max Q Drogue deploy
CDT 3-7	12/20/12	PTV	25 kft	-	1	3	3	Drogue fail
CDT 3-8	02/12/13	PCDTV	25 kft	3	2	3	3	High Q Drogue Deploy, Drogue skip 1st, flagging Main
CDT 3-9	05/01/13	PTV	25 kft	-	1	3	3	Drogue fail, Main skip 1st
CDT 3-11	07/24/13	PTV	35 kft	-	2	3	3	Main skip 1st & released
CDT 3-10	01/16/14	PTV	25 kft	3	2	3	3	FBC & nominal system
CDT 3-12	02/26/14	PCDTV	35 kft	3	2	2	2	Max Q Drogue deploy, Pilot & Main fail
CDT 3-13	04/23/14	PTV	13 kft	-	-	3	3	Straight to Mains deploy
CDT 3-14	06/25/14	PTV	35 kft	3	2	3	3	FBC & Main skip 2nd
CDT 3-15	12/18/14	PTV	25 kft	-	2	2	2	Textile risers, Main design changes
CDT 3-16	8/26/15	PTV	35 kft	2	1	2	2	Minimum System, textile risers, 85% PRL
CDT 3-17	1/13/16	PCDTV	30 kft	2	2	3	3	High Q Drogue and Main deploy

Parachute Qualification Drop Testing

Number	Date	Vehicle	Alt.	Parachute # in Cluster				Primary Test Objective(s)
				FBC	Drogue	Pilot	Main	
CQT 4-1	9/30/16	PCDTV	35 kft	2	2	3	3	Two FBCP, nominal system, bounding high Q Drogue and Main deploys
CQT 4-2	3/8/17	PTV	25 kft	-	2	3	3	No FBCPs, min Q Drogue deploy
CQT 4-3	6/14/17	PTV	25 kft	-	-	3	3	Straight to Mains, low Q deploy
CQT 4-4	9/13/17	PTV	25 kft	-	-	3	3	Straight to Mains, high Q deploy
CQT 4-5	12/15/17	PTV	35 kft	2	2	2	2	Two FBCPs, two Mains
CQT 4-6	3/16/18	PTV	35 kft	2	2	3	3	Nominal system with FBC
CQT 4-7	7/12/18	PCDTV	35 kft	2	2	2	2	Two Pilots/Mains, high Q Drogue and Main deploy
CQT 4-8	9/13/18	PTV	35 kft	3	2	3	3	Nominal system with FBC

▲ Testing at NASA's Langley Research Center in Hampton, Virginia takes place to investigate the heating of the Orion spacecraft during reentry into Earth's atmosphere with a 6-inch Orion heat shield model in the 20-inch Mach 6 wind tunnel.

Wind Tunnel Testing

In order to successfully carry out its various missions, Orion's flight behavior in Earth's atmosphere must be accurately designed and understood. Wind tunnel testing and simulations have played an important role in developing the aerodynamic, aerothermal, and aeroacoustics databases for atmospheric flight of Orion. The databases help to verify the performance, controllability, thermal protection system, structure, and safety of the vehicle during all phases of atmospheric flight, including launch aborts, by allowing accurate flight simulations and informing good design for the vehicle.

Defining the crew module aerodynamics, both static and dynamic, is important in order to ensure stable and controllable flight from entry into Earth's atmosphere to parachute deployment and descent. It is also important to define for the launch abort system (LAS) to ensure successful launch aborts during ascent from the launch pad to orbit. Defining the crew module and LAS aerothermal environments is important to design thermal protection systems that will protect them from heat during atmospheric entry, ascent, and ascent aborts. Characterizing the aeroacoustics is also important to design and test the vehicle structures for the vibrations and loads they will experience during ascent and entry.

The Orion aerosciences testing team has completed more than 120 tests as part of developing the aerodynamic, aerothermal and aeroacoustic databases for Orion. Tests have been conducted in 25 different wind tunnels, 4 ballistic ranges, 2 shock tunnels, and 3 research laboratories across the U.S. The tests have covered a Mach number range of 0.05 to ~20 (38 mph to about 15,000 mph). Tests have been performed at NASA facilities in Virginia, California and Ohio; Department of Defense facilities in Tennessee, Maryland and Florida; and universities such as the University of Buffalo in New York.

During testing, the conditions experienced by the LAS and the crew module are simulated inside the test facility. For tests of ascent aborts, plumes from the LAS motors are simulated with a simulant gas. Air, helium, and rocket fuel were used during these tests to understand the jet interactions of the plumes on the vehicle from aerodynamic, aeroacoustic, and aerothermal perspectives. Tests with the LAS attitude control motor, abort motor, jettison motor, and crew module reaction control systems have all been performed. These data are then analyzed, developed into databases, and aerosciences customers will use those databases to design the flight control systems, thermal protection systems, and structural design of the Orion vehicles.

Wind Tunnel Testing

Test Number	Type	Date	Facility	Description
50-AS	Ascent Acoustics	5/17/07	Boeing Polysonic Wind Tunnel (PSWT)	A preliminary investigation into the aeroacoustic loads generated by the Pad Abort Test (PA-1) LAV configuration and the potential reduction in those loads provided by an alternate Launch Abort System configuration (ALAS-2 mod-1) developed by the ALAS project of the NESC.
58-AA	Ascent Acoustics	10/8/07	Arnold Engineering and Development Center (AEDC) 4T	Test to identify LAV configuration to adopt for flight. Downselect between ALAS-11 rev 3, rev 8, and rev 10. Approximately12 flush mounted microphones
57-AS	Ascent Acoustics	11/1/07	NASA Glenn Research Center (GRC) 8x6	LAV Ascent Aeroacoustics comparing PA-1 and ALAS-11 rev. 3 configurations with approximately 100 surface mounted microphones
11-CD	Dynamic Stability	4/8/06	US Army Aberdeen Test Range	Proof of concept test to evaluate the Aberdeen Research Laboratory telemetry technique for ballistic range test data acquisition and analysis of CM flight.
8-CD	Dynamic Stability	5/10/06	NASA Langley Research Center (LaRC) Transonic Dynamics Tunnel (TDT)	Small amplitude forced oscillation test of CM w/ some unsteady pressures.
12-CD	Dynamic Stability	6/19/06	US Army Aberdeen Test Range	Evaluation of improved sabot designs for CM testing at the Aberdeen Test Range
13-CD	Dynamic Stability	7/6/06	US Air Force Eglin Ballistic Range	Transonic and supersonic dynamic aero data for zero L/D CM model
15-CD	Dynamic Stability	9/2/06	US Army Aberdeen Test Range	Lifting and non-lifting CM models for dynamic aero database development
14-CD	Dynamic Stability	10/5/06	NASA Ames Research Center Hypersonic (ARC) Free-Flight Aerodynamics Facility	Transonic and supersonic dynamic aero data of CM at non-zero L/D
18-CD	Dynamic Stability	1/1/07	LaRC TDT	Demonstration of Oscillating Turn Table test technique in the TDT to obtain dynamic stability of the CM at high Reynolds numbers. Comparisons with ballistic range data and previous small amplitude forced oscillation test (8-CD)
48-CD	Dynamic Stability	3/1/07	LaRC Vertical Spin Tunnel (VST)	Free-flight test of CM at low Mach number to provide dynamic stability estimates for the Pad Abort flight test.
52-CD	Dynamic Stability	3/1/07	ARC Fluid Mechanics Laboratory Test Cell 2 (TC-2)	Test technique development to examine issues related to Free-to-Oscillate testing. Will duplicate the conditions of 48-CD test of the CM.
45-AD	Dynamic Stability	3/9/07	LaRC VST	Low-Mach number test of LAV in support of PA-1 Flight Test
29-CD	Dynamic Stability	6/1/07	ARC Gun Development Facility	Phase 2 of CM dynamic stability at large angles of attack.
82-AD	Dynamic Stability	12/21/07	LaRC VST	Forced Oscillation test of LAV in the Vertical Spin Tunnel

Test Number	Type	Date	Facility	Description
27-AD	Dynamic Stability	3/21/08	LaRC TDT	Forced oscillation (subsonic and transonic) test of LAV and CM through as much of the 0-180 deg. range as possible
108-CD	Dynamic Stability	8/25/09	Bihrle Research VST	Low-speed dynamic stability test to support Orion decisions on back shell angle changes.
109-CD	Dynamic Stability	11/15/09	09 LaRC VST	Dynamic stability of CM under parachutes.
117-CD	Dynamic Stability	4/1/10	LaRC VST	Phase 2 of dynamic stability test of CM under parachute
46-AD	Dynamic Stability	6/1/10	US Air Force Eglin Ballistic Range	Ballistic range test of LAV.
55-AS	Plume Acoustics	9/28/07	Florida State Jet Noise Laboratory	Series of hot- versus cold-jet acoustic experiments to possibility develop scaling laws to allow the use of cold plume tests for the LAV with AM firing. Phase 1 - 2" 2,000°F jet versus 2" cold jet. Phase 2 - Same 2 jets with more measurement locations. Phase 3 - ~1" D nozzle exit hybrid rocket. Phase 4 - Sounding rocket motor plume noise measurements at NASA Wallops Flight Facility.
51-AS	Plume Acoustics	10/30/0	ARC Unitary Plan Wind Tunnel (UPWT)	6%-scale LAV model test to determine the aeroacoustic loading generated by cold air simulation of the AM plumes. ~200 flush microphones.
80-AS	Plume Acoustics	9/20/10	ARC UPWT	Hot Helium simulation of AM plumes for acoustic loads. ~200 flush microphones.
53-AA	Plume Jet Interaction (JI)	5/21/07	Texas A&M 7x10 Foot Wind Tunnel	First test of subsonic interactions between the ACM plumes and the LAV. Primarily to validate CFD and to provide some data on coast-phase ACM increments for the PA-1 flight test aero database.
16-AA	Plume JI	6/22/07	ARC UPWT	Abort loads on the CM due to AM plume JI and proximity to Service Module
59-AA	Plume JI	9/14/07	ARC UPWT	High fidelity ACM JI for both Pad Abort-1 flight test article and production ALAS-11rev3B configuration.
60-AA	Plume JI	1/30/08	ARC UPWT	Preliminary separation aerodynamics during abort initiation on PA-1 and ALAS-11 rev3B configurations. Preliminary aeroacoustics for nominal ascent (with SM) and abort (LAV only) with cold air plume simulation.
85-AA	Plume JI	8/20/08	GRC Aero-Acoustic Propulsion Laboratory	CFD validation test documenting flowfield associated with single AM nozzle at M < 0.3 using PIV. Nozzle at 0°, 25°, and 40° relative to free stream. With and without simplified LAV model.
61-AA	Plume JI	12/1/08	LaRC 14- by 22 Foot Wind Tunnel	Subsonic 6%-scale Jettison Motor Jet Interaction test around alpha = 180°.
24-AA	Plume JI	6/25/09	AEDC 16T	Transonic/supersonic test of Jettison Motor Jet Interation for LAS jettison during a launch abort (i.e. heat shield forward).
75-AA	Plume JI	7/24/09	ARC UPWT	Subsonic, transonic, and low-supersonic ACM Jet Interaction test.
76-AA	Plume JI	11/24/0	LaRC UPWT	Supersonic ACM Jet Interaction test (M 1.6 to 4.6).
25-AA	Plume JI	2/26/10	ARC UPWT	Supersonic (M 1.6 to 2.5) Jettison Motor Jet Interaction and Jettison LAS/CM Proximity aerodynamics.
26-AA	Plume JI	8/9/10	ARC UPWT	Subsonic, transonic, and supersonic AM and ACM Jet Interactions including separation effects data for the LAV. PSP to document pressure loadings during launch aborts.

Test Number	Type	Date	Facility	Description
3-CA	CM Static Aero	2/10/06	LaRC UPWT	Study of BL trip techniques on 3%-scale CM model.
7-CA	CM Static Aero	3/10/06	LaRC UPWT	Force and moment measurements & pressure distributions, with apex cover on/off and boundary layer transition/tripping study.
5-CA	CM Static Aero	3/22/06	ARC UPWT	Force & moments and pressure data on 7.5%- and 3%-scale models. Provided tunnel-to-tunnel comparisons between LaRC and ARC UPWT.
9-CA	CM Static Aero	4/20/06	LaRC Mach 6 Tunnel	3%-scale CM test for alpha from 0 to 180°.
1-CA	CM Static Aero	12/8/06	LaRC UPWT	Boundary layer transition measurements with IR thermography and Temperature Sensitive Paint.
19-AA	LAV Static Aero	1/29/07	Boeing PSWT	3%-scale transonic test of PA-1 LAV configuration for 0-180° angle of attack.
54-AA	LAV Static Aero	4/13/07	Lockheed High-Speed Wind Tunnel	Quantify the roll coupling with angle of attack caused by the clocking of the abort motor nozzles on the PA-1 configuration. Study effectiveness of various nozzle fairings in relieving the roll interaction
88-AA	LAV Static Aero	10/20/0	LaRC VST	Test of the PA-1 Launch Abort Tower alone to define the postjettison aerodynamics.
83-AA	LAV Static Aero	6/1/08	LaRC National Transonic Facility	High-Re effects on unpowered LAV aerodynamics.
122-PA	Static Aero	11/8/10	ARC TC-2	Small-scale test of Forward Bay Cover aerodynamics to validate CFD and engineering models of the FBC jettison event.

Final Operations and Assembly at Kennedy

Operations and Checkout Building

The Neil Armstrong Operations & Checkout (O&C) Building has played a major role in NASA's spaceflight history. It was the first building finished at NASA's Kennedy Space Center and has housed the Astronaut Crew Quarters since the mid-1960s where the Gemini astronauts stayed prior to launch.

When the Orion spacecraft arrives to Kennedy following initial manufacturing at NASA's Michoud Assembly Facility in New Orleans, it is placed in the O&C Building. The O&C contains a large room called a high bay that is operated by Lockheed Martin as a high-tech factory; it is where the spacecraft is assembled and readied for missions to deep space destinations. The high bay underwent an extensive, two-year renovation starting in 2007 to outfit the facility for assembly of Orion.

It includes unique tooling stations, test fixtures, chambers and clean rooms for the buildup and testing of the spacecraft and two 27.5-ton cranes used to move payloads.

Orion assembly in the O&C Building consists of joining the crew module, crew module adapter, and the service module which will carry the consumables astronauts will need for missions into deep space. Various testing on the crew and service modules also takes place in this facility, including a final round of testing and assembly in the Final Assembly and System Testing (FAST) cell. After final checkout in the O&C, the spacecraft is then transported to the Multi-Payload Processing Facility (MPPF) and handed over to Exploration Ground Systems for ground processing.

▼   A variety of space hardware for the Orion crew module is in view in the Neil Armstrong Operations and Checkout Building high bay at NASA's Kennedy Space Center in Florida.

11

The 19,647-square-foot Multi-Payload Processing Facility at Kennedy Space Center in Florida is where Orion will receive its flight load of propellant, high pressure gasses, and coolant.

Multi-Payload Processing Facility

A unique facility at Kennedy, the Multi-Payload Processing Facility (MPPF) is used for commodity servicing and fueling the Orion spacecraft with its flight load of hazardous propellants, high pressure gases, coolant, and other fluids the spacecraft and astronauts will need to maneuver and carry out their missions in space. When Orion returns to Earth after its mission, it will be transported back to the MPPF, where specialized equipment will be used to remove unused hazardous propellants from its tanks during spacecraft postflight processing.

On Jan. 16, 2021, the Artemis I vehicle was transferred from the Operations & Checkout (O&C) Building to the MPPF, where it was moved onto a service stand that provides 360-degree access, allowing engineers and technicians from Exploration Ground Systems, its prime contractor Jacobs Technology, and other support organizations to fuel and service the spacecraft. Crane operators removed the transportation cover and used fuel lines and several fluid ground support equipment panels to load the various commodities into the crew and service modules.

Both pieces of hardware underwent fueling and commodity servicing in the facility ahead of launch. The ICPS will be positioned above the core stage and will provide the power needed to give Orion the big push it needs to break out of Earth orbit on a precise trajectory toward the Moon during Artemis I.

The fueling of hazardous commodities is performed remotely from a firing room in the Launch Control Center. Various electrical ground support equipment racks allow technicians to power up the spacecraft and perform service operations remotely. The spacecraft's temperature and humidity are tightly controlled using mini-portable purge units, which provide a constant flow of conditioned air.

The 19,647-square-foot MPPF building, originally constructed in 1995, has undergone extensive upgrades and modernizations to support Orion, which began in 2013. The facility also will be used to install time-critical crew equipment and to process crew modules that have returned from space prior to being returned to the O&C for possible reuse. Ground support equipment also will be stored and maintained in the MPPF.

After Orion is fueled and final checks are performed in the MPPF, its transportation cover is re-installed, and the spacecraft is moved to the Launch Abort System Facility (LASF) for integration with the launch abort system.

Launch Abort System Facility

After fueling and commodity servicing in the MPPF, the Orion spacecraft is moved to the Launch Abort System Facility (LASF). The 44-foot-tall launch abort system (LAS) is first prepared horizontally inside the LASF and is afterwards assembled vertically with the Orion spacecraft. The LASF, previously used to lift and rotate the canister that carried space shuttle payloads to the launch pad, is taller than many processing facilities at Kennedy to allow clearance for vertical assembly and has cranes and other equipment needed to integrate the system during launch processing. Installation of the ogive panels that protect the crew module and LAS and provide its aerodynamic shape also take place at the LASF. After the LAS is integrated with Orion, the entire stack is moved to High Bay 3 in the Vehicle Assembly Building. The Artemis I spacecraft moved into the LASF for integration on July 21, 2021.

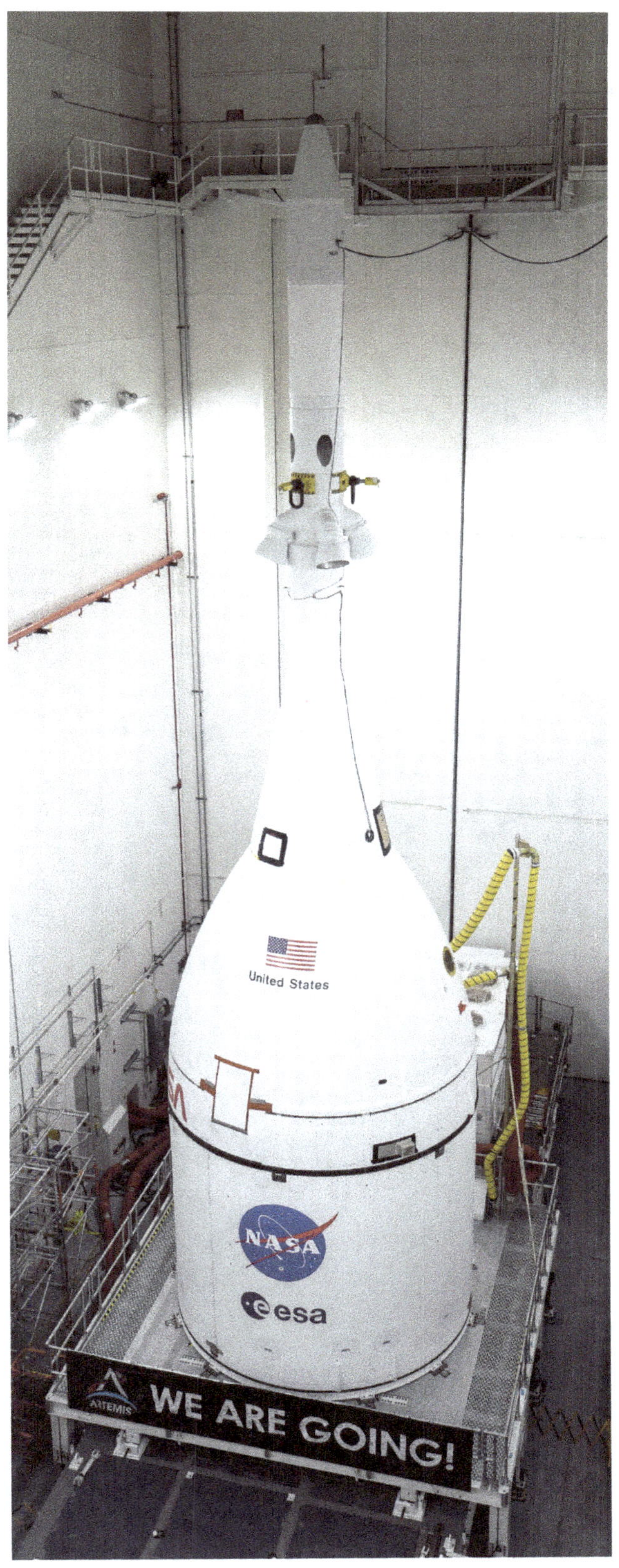

▶ Inside the Launch Abort System Facility (LASF) at NASA's Kennedy Space Center in Florida, the fully assembled Orion spacecraft for the agency's Artemis I mission is prepared for transport on Oct. 18, 2021.

Vehicle Assembly Building

The Vehicle Assembly Building (VAB) serves as the central hub of NASA's premier multi-user spaceport, capable of hosting several different kinds of rockets and spacecraft at the same time. Whether the rockets and spacecraft are going into Earth orbit or being sent into deep space, the VAB has the infrastructure to prepare them for their missions.

The tallest portion of the VAB is called the high bay. There are four high bays, two on the east side, and two on the west side of the building. Each has a 456-foot-high door, enabling rockets to be stacked vertically and then rolled out to the launch pad.

The massive SLS rocket and its twin solid rocket boosters require a large high bay from which to process and prepare them for launch. Exploration Ground Systems recently completed many upgrades and modifications to High Bay 3 in the VAB where the SLS and Orion are stacked, processed and tested atop the new mobile launcher. Ten levels of new work platforms were installed in High Bay 3, Towers E and F, to accommodate the rocket. A total of 20 platform halves altogether surrounds the SLS and Orion spacecraft and provide access for testing and processing before they are transported to the launch pad.

▼ Seen to the right of the iconic Vehicle Assembly Building at NASA's Kennedy Space Center in Florida, a crane positions the Orion crew access arm (CAA) so it can be attached to the mobile launcher (ML).

Each of the giant steel platforms measures about 38 feet long and close to 62 feet wide. Each weigh between 300,000 and 325,000 pounds. The platforms are attached to rail beams which provide structural support and contain the drive mechanisms to retract and extend them. Each platform rides on four Hillman roller systems that are located two on each side – much like a kitchen drawer glides in and out. A mechanical articulated tray also moves in and out with each platform.

In the VAB, the Orion spacecraft is lifted and secured atop the SLS rocket on the mobile launcher. The completed SLS/Orion stack is then transported atop the mobile launcher on Crawler-Transporter 2 to Launch Pad 39B for launch.

The Artemis I Orion spacecraft was moved from the Launch Abort System Facility (LASF) to the VAB on Oct. 19, 2021. The Orion spacecraft and SLS rocket were hard mated on Oct. 21, 2021, marking completion of the Artemis I stack.

Inside the Vehicle Assembly Building at NASA's Kennedy Space Center in Florida, a crane lowers the Space Launch System (SLS) core stage pathfinder into High Bay 3 on Oct. 16, 2019.

Resources

Acronyms

AA-2	Ascent Abort-2
ABCU	Auxiliary Bus Control Unit
ACM	Attitude Control Motor
ADP	Advanced Development Projects
AFRC	Armstrong Flight Research Center
AI&P	Assembly, Integration and Production
AI&T	Analysis, Integration & Test
AM	Abort Motor
AOS	Acquisition of Signal
APS	Avionics Power and Software
ARC	Ames Research Center
ATB	Abort Test Booster
ATCS	Active Thermal Control System
ATD	Anthropomorphic Test Devices
ATLO	Assembly Test & Launch Operations
ATP	Authority To Proceed
ATV	Automated Transfer Vehicle
BEO	Beyond Earth Orbit
BTA	Boilerplate Test Article
C&DH	Command & Data Handling
C&T	Communications & Tracking
CAM	Control Account Manager
CCB	Configuration Control Board
CDR	Critical Design Review
CE	Chief Engineer
CM	Crew Module
CMA	Crew Module Adapter
CMO	Chief Medical Officer
CMRT	Crew Module Recovery Trainer
CMUS	Crew Module Uprighting System
COFR	Certification of Flight Readiness
CPIT	Cross Program Integration Team
CSM	Crew & Service Module
CSO	Chief Safety Officer
CT	Crawler Transporter
CTIL	Communication & Tracking Integrated Lab
D&C	Display & Controls
DCR	Design Certification Review
DD250	Department of Defense form 250
DDT&E	Design, Development, Test & Evaluation
DFI	Developmental Flight Instrumentation
DM	Data Management
DMJS	Docking Mechanism Jettison System
DOD	Department of Defense
DU	Display Unit
ECB	Exploration Systems Development Control Board
ECD	Estimated Completion Date
ECLS	Environmental Control and Life Support
ECLSS	Environmental Control and Life Support System
ECN	Engineering Change Notice
EDL	Exploration Development Lab
EDU	Engineering Development Unit
EEE	Electrical, Electronic and Electromechanical
EERB	ESD Engineering Review Board
EFT-1	Exploration Flight Test 1
EGF	Flight Support Equipment Grapple Fixture
EGS	Exploration Ground Systems
EGSE	Electrical Ground Support Equipment
EIO	European Service Module Integration Office
EMC	Electromagnetic Compatibility
EMI	Electromagnetic Interference
ERB	Engineering Review Board
ESA	European Space Agency
ESD	Exploration Systems Development
ESD CE	Exploration Systems Development Chief Engineer
ESD RMO	Exploration Systems Development Resources Management Office
ESM	European Service Module
E-STA	European Structural Test Article

EUS	Exploration Upper Stage	ICPS	Interim Cryogenic Upper Stage	MECO	Main Engine Cut Off
EVA	Extra-vehicular Activity	ICWG	Interface Control Working Group	MET	Mission Elapsed Time
FAST	Final Assembly and Test			MEV&V	Multi-Element Verification & Validation
FBC	Forward Bay Cover	IG	Inspector General		
FBCP	Forward Bay Cover Parachutes	IML	Inner Mold Line	MGSE	Mechanical Ground Support Equipment
FCR	Flight Control Room	IOZ	Industrial Operating Zone		
FDR	Flight Data Recorder	IPIT	Integrated Programmatic and Program Control Integration Team	MIR	Mission Integration Review
FIT	Focused Integration Team			ML	Mobile Launcher
FMEA/CIL	Failure Modes Effects Analysis/ Critical Items List			MLI	Multi-layer insulation
		IPO	Initial Power On	MMH	Monomethylhydrazine (hypergolic propellant)
FOD	Flight Operations Directorate	IPT	Integrated Product Team		
FOP	Flight Operations Teams	IRF	Integrated Robotics Facility	MMOD	Micro Meteor Orbital Debris
FPCB	ESD Flight Procedures Control Board	ISS	International Space Station	MMT	Mission Management Team
		ITAR	International Traffic in Arms Regulation	MOD	Mission Operations Directorate
FRCB	ESD Flight Rules Control Board			MODE	Multi-Organizational Development Engineering
FRR	Flight Readiness Review	ITL	Integrated Test Laboratory		
FSW	Flight Software	IWG	Integrated Working Group	MON	Mixed Oxides of Nitrogen
FTV	Flight Test Vehicle	JICB	Joint Integration Control Board	MPCB	MPCV Program Control Board
FY	Fiscal Year	JM	Jettison Motor	MPCV	Multi-Purpose Crew Vehicle
GC	Ground Controller	JPCB	Joint Program Control Board	MPPF	Multi-Payload Processing Facility
GEMCB	Government Equipment & Materials Control Board	JSC	Johnson Space Center		
		KDR	Key Driving Requirement	MSERP	MPCV Safety Engineering Review Panel
GFE	Government Furnished Equipment	KSC	Kennedy Space Center		
		LaRC	Langley Research Center	MSFC	Marshall Space Flight Center
GMIP	Government Mandatory Inspection Point	LAS	Launch Abort System	MSO	Mass Simulator for Orion
		LASF	Launch Abort System Facility	M-STA	Mechanical Structural Test Article
GMT	Greenwich Mean Time	LCCP	Cross Program Launch Commit Criteria Panel		
GNC	Guidance, Navigation & Control			MVF	Mechanical Vibration Facility
GOOV	Gas On/Off Valve	LEAF	Lunar Environment Arcjet Facility	MVS	Mechanical Vibration System
GPS	Global Positioning System	LEO	Low-Earth Orbit	NBL	Neutral Buoyancy Laboratory
GRC	Glenn Research Center	LETF	Launch Equipment Test Facility	NPD	NASA Policy Directive
GSE	Ground Support Equipment	LIDAR	Light Detection and Ranging	NPR	NASA Program Requirement
GTA	Ground Test Article	LIDS	Low Impact Docking System	O&C	Operations and Checkout
HEO	Human Exploration and Operations Mission Directorate	LM	Lockheed Martin	OCHMO	Office of the Chief Health and Medical Officer
		LOC	Loss of Crew		
HH&P	Human Health & Performance	LOI	Lunar Orbit Insertion	OCOMM	Office of Communications
HITL	Human in the Loop	LOM	Loss of Mission	ODN	Onboard Data Network
HLS	Human Landing System	LOS	Loss of Signal	OERB	Operations Engineering Review Board
HMTA	Health & Medical Technical Authority	LOX	Liquid Oxygen		
		LRD	Launch Readiness Date	OFT	Orion Flight Test
HOTH	Houston Orion Test Hardware	LRS	Landing & Recovery System	OIMU	Orion Inertial Measurement Unit
HS	Heat Shield	M&P	Materials and Processes	OMB	Office of Management & Budget
HW	Hardware	MAF	Michoud Assembly Facility	OML	Outer Mold Line
IBR	Integrated Baseline Review	MCC	Mission Control Center		

OMS-E	Orbital Maneuvering System Engine	QPSR	Quarterly Program Status Review
OMSRP	Operations & Maintenance Requirements & Specifications Panel	R&R	Release & Retention
		R2D2	Rapid Reconfigurable Data Distribution
OPOC	Orion Production and Operations Contract	RATF	Reverberant Acoustic Test Facility
OPSR	Orion Program Schedule Review	RCS	Reaction Control System
ORR	Operational Readiness Review	RFP	Request for Proposal
OSMA	Office of Safety and Mission Assurance	ROC	Reconfigurable Operational Cockpit
OSMS	Orion Summary Master Schedule	RPL	Rapid Prototyping Laboratory
OVIT	Orion Vehicle Integration Team	RPOD	Rendezvous, Prox ops and Docking
PA-1	Pad Abort-1	RPSF	Rotation, Processing & Surge Facility
PAA	Phased Array Antenna		
PBS	Plum Brook Station	S&MA	Safety and Mission Assurance
PCA	Pressure Control Assembly	SA	Spacecraft Adapter
PCB	Program Control Board	SADE	Solar Array Drive Electronics
PCDTV	Parachute Compartment Drop Test Vehicle	SADM	Solar Array Drive Mechanism
PCDU	Power Conditioning and Distribution Unit	SAJ	Spacecraft Adapter Jettison
		SAR	System Acceptance Review
PDE	Propulsion Drive Electronics	SAT	System Acceptance Test
PDR	Preliminary Design Review	SAW	Solar Array Wings
PDU	Power and Data Unit	SCAN	Spacecraft Communications & Navigation
PLE	Project Lead Engineer		
PMRB	Program Material Review Board	SCAPE	Self-Contained Atmospheric Protective Ensemble
PORT	Post-Landing Orion Recovery Tests	SE&I	Systems Engineering & Integration
PPBE	Planning, Programming, Budgeting, and Excecution	SEC	Space Environments Complex
		SEQR	Senior Executive Quarterly Review
PQM	Propulsion Qualification Module	SET	Space Environmental Test (Facility)
PRA	Probabalistic Risk Assessment		
PRU	Pressure Regulator Unit	SET	Space Environment Test Facility
PSMS	Program Master Summary Schedule	SICB	SLS Integration Control Board
PTC	Passive Thermal Control	SIM	Simulation
PTV	Parachute Test Vehicle	SLF	Spacecraft Leadership Forum
PV	Pressure Vessel	SLS	Space Launch System
QPM	Quarterly Progress Meeting (Airbus)	SM	Service Module
		SM&A	Safety Mission & Assurance
QPPR	Quarterly Program Performance Review	SMCB	Service Module Control Board

SNIC	Standard Network Interface Card
SPLASH	Structural Passive Landing Attenuation for Survivability of Human Crew (Project)
SPT	Subsystem Product Teams
SRA	Schedule Risk Analysis
SRR	System Requirements Review
SSDL	Support Software Development Lab
SSM	Subsystem Manager
STA	Structural Test Article
STORRM	Sensor Test for Orion Relative Navigation Risk Mitigation
SVMF	Space Vehicle Mock-up Facility
SW	Software
T&V	Test & Verification
TBR	To Be Resolved
TCU	Thermal Control Unit
TEI	Trans-Earth Injection
TI	Terminal Initiation
TLI	Trans-Lunar Insertion
TPR	Top Program Risk
TPS	Thermal Protection System
TRL	Technology Readiness Level
TVAC	Thermal Vacuum Chamber
TVC	Thrust Vector Control
VAB	Vehicle Assembly Building
VAC	Verification Analysis Cycle
VICB	Vehicle Integration Control Board
VIO	Vehicle Integration Office
VMC	Vehicle Management Computer
VNS	Vision Navigation Sensor
VPU	Video Processing Unit
VTB	Verification Test Bed
VVM	Visiting Vehicle Manager
WG	Working Group
WIT	Water Impact Testing
WOOV	Water On/Off Valve
WSTF	White Sands Test Facility

For more information about the Orion program, visit
https://www.nasa.gov/orion

https://www.facebook.com/NASAOrion

https://www.twitter.com/NASA_Orion
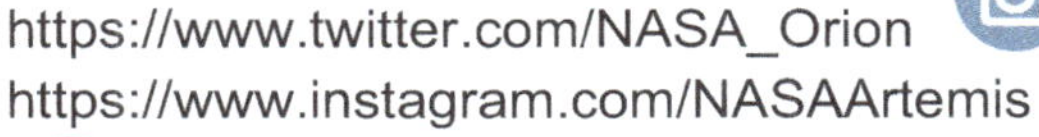
https://www.instagram.com/NASAArtemis

https://www.flickr.com/photos/nasaorion/

Email: w2tap@arrl.net

National Aeronautics and Space Administration

NASA Johnson Space Center
1 E NASA Pkwy
Houston, TX 77058

ww.nasa.gov